THE SHAPE & FORM OF PUGET SOUND

THE SHAPE & FORM OF PUGET SOUND

Robert Burns

A Washington Sea Grant Publication
Distributed by the University of Washington Press
Seattle and London

First published in 1985 by
Washington Sea Grant Program
University of Washington

Distributed by University of Washington Press
Seattle, Washington 98195

Printed in the United States of America

Publication of this book was supported by grants (04-5-158-48; 04-7-158; NA79AA-D-00054, NA81AA-D-00030, and NA84AA-D-00011) from the National Oceanic and Atmospheric Administration and by funds from the Environmental Protection Agency. Writing and publication was conducted by the Washington Sea Grant Program under project A/PC-7.

Library of Congress Cataloging in Publication Data

Burns, Robert E., 1925-
The shape and form of Puget Sound.
(Puget Sound Books) (A Washington Sea Grant Publication)
Bibliography: p.
Includes index.
1. Geomorphology—Washington (State)—Puget Sound
Region. 2. Geology—Washington (State)—Puget Sound
Region. 3. Oceanography—Puget Sound (Wash.).
I. Title. II. Series. III. Series: Washington Sea Grant publication.
GB458.8.B87 1985 551.46′632 84-15354
ISBN 0-295-96184-8

Contents

To Phoebe . . .
To Emily . . .
To Alice. . .
Three generations who share my love for Puget Sound.

Funds to support the publication of the Puget Sound Books were provided by the National Oceanic and Atmospheric Administration (NOAA) and by the Environmental Protection Agency (EPA).

About the Puget Sound Books

This book is one of a series of books that have been commissioned to provide readers with useful information about Puget Sound . . .

About its physical properties—the shape and form of the Sound, the physical and chemical nature of its waters, and the interaction of these waters with the surrounding shorelines.

About the biological aspects of the Sound—the plankton that form the basis of its food chains; the fishes that swim in this inland sea; the region's marine birds and mammals; and the habitats that nourish and protect its wildlife.

About man's uses of the Sound—his harvest of finfish, shellfish, and even seaweed; the transport of people and goods on these crowded waters; and the pursuit of recreation and esthetic fulfillment in this marine setting.

About man and his relationships to this region—the characteristics of the populations which surround Puget Sound; the governance of man's activities and the management of the region's natural resources; and finally, the historical uses of this magnificent resource—Puget Sound.

To produce these books has required more than six years and the dedicated efforts of more than one hundred people. This series was initiated in 1977 through a survey of several hundred potential readers with diverse and wide-ranging interests.

The collective preferences of these individuals became the standards against which the project staff and the editorial board determined the scope of each volume and decided upon the style and kind of presentation appropriate for the series.

In the Spring of 1978, a prospectus outlining these criteria and inviting expressions of interest in writing any one of the volumes was distributed to individuals, institutions, and organizations throughout Western Washington. The responses were gratifying. For each volume no fewer than two and as many as eight outlines were submitted for consideration by the staff and the editorial board. The authors who were subsequently chosen were selected not only for their expertise in a particular field but also for their ability to convey information in the manner requested.

Nevertheless, each book has a distinct flavor—the result of each author's style and demands of the subject being written about. Although each volume is part of a series, there has been little desire on the part of the staff to eliminate the individuality of each volume. Indeed, creative yet responsible expression has been encouraged.

This series would not have been undertaken without the substantial support of the Puget Sound Marine EcoSystems Analysis(MESA) Project within the Office of Oceanography and Marine Services/Ocean Assessment Division of the National Oceanic and Atmospheric Administration. From the start, the representatives of this office have supported the conceptual design of this series, the writing, and the production. Financial support for the project was also received from the Environmental Protection Agency and from the Washington Sea Grant Program. All these agencies have supported the series as part of their continuing efforts to provide information that is useful in assessing existing and potential environmental problems, stresses, and uses of Puget Sound.

Any major undertaking such as this series requires the efforts of a great many people. Only the names of those most closely associated with the Puget Sound Books—the writers, the editors, the illustrators and cartographers, the editorial board, the project's administrators and its sponsors—have been listed here. All these people—and many more—have contributed to this series, which is dedicated to the people who live, work, and play on and beside Puget Sound.

Alyn Duxbury and Patricia Peyton
November 1984

Preface

The water surface defines two of Puget Sound's four dimensions. These have been relatively accessible for some time and have played a prominent role in the development of the Pacific Northwest. Less accessible than the surface, water depth is a third dimension. Historically, water depth has had little importance principally because the water is relatively deep over much of the area. This is, in turn, directly related to the processes which have shaped the area and which provide the time dimension.

In discussing the processes which have formed Puget Sound it is difficult to pick the most appropriate starting point. My choice is perhaps older than it might be. The smaller scale details of shape of the seafloor of Puget Sound are most directly related to glacial processes that were active only thousands of years ago. But the way in which these glacial processes were controlled by the spatial arrangement of the mountains and lowlands suggests an earlier date might be desirable. I have chosen the latter alternative and will consider several hundred million years evolution of the western margin of North America.

Robert Burns
November 1984

Acknowledgments

The material covered here is based on information synthesized from far too many sources to be individually acknowledged. Some of these sources are indicated in the Bibliography which, although not intended to be comprehensive, should provide the interested reader with a sampling of available source material.

The evolution of some of my earlier draft manuscripts into the present form has been aided by several patient and helpful readers as well as by a relatively "tough" editorial group at the University of Washington's Sea Grant Office.

Figure 1.1 Drainage basin of Puget Sound, the Strait of Georgia, and the Strait of Juan de Fuca. The area shown represents local drainage; additional freshwater is brought in by the Fraser River from a major drainage basin east of the Coast Mountains of British Columbia (After Environment Canada, Lands Directorate 1973).

CHAPTER ONE

The Setting

At the western edge of North America and the boundary between the United States and Canada, occupying part of a regional topographic depression of the earth's crust, is Puget Sound and its sibling Hood Canal. Extending beyond the shores of Puget Sound, a regional depression creates an interior lowland between the coastal mountains and the Cascades that reaches from British Columbia through central Oregon. Much of this interior lowland is occupied by an inland sea. Although geographic proximity, contiguous boundaries, and shared oceanographic conditions make this inland sea a continuum, historical tradition and local usage separates it into three distinct parts: Strait of Juan de Fuca, Strait of Georgia, and Puget Sound (Figure 1.1).

The Strait of Juan de Fuca connects the Pacific Ocean with the flooded portion of the interior lowland. Within the interior lowland, the Strait of Georgia separates Vancouver Island from the mainland of British Columbia. Similarly, Puget Sound extends to the south, flooding part of Western Washington between the Olympic and Cascade mountains. The shape and size of these inland waters are directly related to regional topographic features that are part of the overall structure of western North America.

Neither the shape nor the structural framework of western North America have been clearly defined until relatively recently in earth's history. Although the native inhabitants of the Pacific Northwest were well aware of both the extensive inland waterways of Puget Sound and the adjacent high mountains, the existence of these features was not immediately obvious to European explorers. These early visitors to the western shores of North America came principally by ship, and they found the Pacific Coast considerably less hospitable than the Atlantic and Gulf coasts.

Unlike the Atlantic Coast with its many rivers, open estuaries, bays, and accessible coastal plain, the Pacific Coast is bordered by mile after mile of high cliffs, has few extensive beaches or sheltered bays, and has a minimal coastal plain. In many places rugged mountains parallel the shore, and sea level access to the interior was not immediately discovered by early explorers. From southern California northward there are only three places where there is a natural, navigable, sea level passage through the coastal mountains. Golden Gate provides access into San Francisco Bay and, via the Sacramento and San Joaquim rivers, the Central Valley of California. Farther north, the Columbia River transects not only the coastal mountains but also provides natural access eastward through the Cascade Mountains. Still farther north, the Strait of Juan de Fuca transects the coastal mountains and connects eastward with the major inland estuarine system of the Strait of Georgia and Puget Sound.

This regional estuarine system forms a transition between several rivers and the Pacific Ocean. In the Straits of Georgia and Juan de Fuca estuarine conditions are influenced by freshwater input from the Fraser River. By far the largest of the rivers in the region, the Fraser drains a basin that extends well to the east of the Coast Mountains of British Columbia. Puget Sound is part of the regional estuarine system, but it receives freshwater input principally from smaller, more localized rivers which, within the Puget Sound drainage basin, are relatively more important than the Fraser River outflow.

Although the Spaniards explored along much of the Pacific Coast during the sixteenth and seventeenth centuries, their efforts appear to have been sporadic probes northward from their base in Mexico. By the late sixteenth century there were poorly documented reports of a "Great River of the West" (which could have been the Columbia) and identification of a northern passage opening eastward into a major inland sea. The latter was reported in 1592 by Apostolos Valerianos, a Greek sailing in service of Spain, and was given his Spanish appellation, Juan de Fuca. Although not generally used for almost another 200 years, Juan de Fuca's name persisted because English Captain Charles Barkley recognized the prior exploration when, in 1787, he logged his ship's entry into the strait.

Both Spain and England explored and mapped the Pacific Coast during the latter half of the eighteenth century. In 1778, on his final voyage, Captain James Cook sailed the *Resolution* northward along the coast but apparently remained far enough offshore so that he observed neither the Columbia River nor the Strait of Juan de Fuca. By 1791, the Spanish Captain Quimper had entered the strait, explored and mapped a part of the region, and left—along with other Spanish place names—his own for the Quimper Peninsula at the west of the entrance to Puget Sound.

A thorough exploration of the Puget Sound region was conducted by Captain George Vancouver, who, among other objectives, was specifically charged to determine the true nature of Juan de Fuca's strait, the extent of the inland waters, and whether a passage existed between the Pacific Ocean and Hudson Bay. Vancouver published his findings in 1798, and, although he did not report discovery of the sought after eastward passage, he provided an accurate description of the shape and extent of the regional estuarine system. He also provided many of the local place names—Vancouver Island, Hood Canal, Mount Rainier, Deception Pass, among others—which remain in use today. Puget Sound was named by Vancouver for one of his lieutenants, Peter Puget, who explored the southern inlets beyond The Narrows.

It is interesting to note that, of the three sea level entrances eastward through the coastal mountains, it was the northernmost which was documented earliest. The southernmost, San Francisco Bay, was actually discovered from land in 1769, and was not entered from the sea until 1775 when Lieutenant Juan Manuel de Ayala traversed the Golden Gate on his ship, the *San Carlos*. The Columbia River was entered independently by Captain Robert Gray of Boston in 1792 while Vancouver was working farther to the north.

In this book, the shape and form of Puget Sound will be examined: how it was formed and the processes responsible for shaping it, its present submarine features, and the ongoing changes in its shape and the processes that cause them. Although our attention will focus on Puget Sound, the scale of many of the processes involved is large enough that the regional lowland, the adjoining uplands of the Olympics and Cascades and even larger scale features of the earth as a whole will be examined.

CHAPTER TWO

Mapping the Earth's Surface

About 70 percent of the earth's surface is a relatively flat and featureless ocean; but where continents and oceanic islands project above the ocean's surface, a variety of shapes take form. Early maps showed relative positions, distances, and directions of points of interest, yet because land surface is not a simple plane, direction and distance between two points do not always represent the shortest routes. Consequently, mappers began to include landforms, prompted by man's need to determine the easiest way to get from place to place.

From space, the earth appears to be a smooth sphere, but a closer inspection reveals ups and downs, mountains and valleys, continents and oceans, all of which represent small changes in distance from the earth's surface to its center. To map the shape of this surface, early mapmakers—since they couldn't measure distances between the center of the earth and points at the surface—determined distances of the land surface above sea level. This was both convenient and practical and proved to be a good choice since the surface of the ocean (sea level) provides a reasonably close approximation of a surface that is always the same distance from the center of the earth. Although the ocean's surface is distorted slightly by the earth's rotation and the tidal attraction of the moon and sun, it is the closest naturally occurring approximation of a theoretical spherical surface. As the art of mapping developed, sea level remained a basic reference surface. On land, because distances from the center of the earth vary, they are reported as elevations above sea level or as depths below sea level.

The general sizes, shapes, and spatial relationships of land masses and oceans were mapped as early as the sixteenth century, but not with reasonable accuracy until the middle of the nineteenth century. On land, definition of shapes permitted identification and classification of landforms over a large portion of the known world. By the beginning of the twentieth century there were few surprises left about the shape of the land surface and by the 1950s much of it had been mapped in reasonable detail using aerial photography. During the 1970s, the advent of satellite imagery left very little of the land's surface that had not been observed, measured, and mapped.

The shape of the seafloor is as varied as the land, but mapping of submarine landforms has been considerably slower to develop than mapping on land. Although landforms can easily be seen by eye and measured by aerial photography and satellite imagery, seafloor forms are not so easy to see from the water's surface. However, for a variety of reasons, most of them practical, the question of how deep the water is became important to society many years ago. Initially, prompted by problems associated with safe navigation for shipping, topographic mapping was expanded to the nearshore waters and included identification of rocks and shoals as well as delineation of safe channels.

Early techniques for measuring depths below sea level were relatively simple; a rod, pole, or weight at the end of a line to feel for the bottom and measure depth. Using these primitive techniques information was slowly accumulated that provided insight into the shape of the seafloor in the shallow water nearshore.

In deeper water the probability of dangerous shoals was considerably lessened and soundings were taken much less frequently. Sounding with weight and line was very time consuming; the ship had to stop for an extended period while the weight was lowered, the bottom felt, the depth of water recorded, and the weight retrieved. The motivation for sounding in deep water was principally scientific through the early part of the twentieth century and, consequently, relatively little information about the deep seafloor was assembled other than a general indication of the existence of some prominent submarine topographic features.

The use of sound for determining water depth was developed during the First World War and became more common in the years that followed. Although seawater is opaque to light, it is transparent to sound. Echo-sounding involves measuring the time required for a sound wave originated at a ship on the sea surface to travel downward through the water, reflect off the seafloor, and return to the surface. The depth of the water is then calculated from the travel time of the sound waves and the speed of sound through seawater.

Echo-sounding provided a much faster and more efficient way to determine water depth and it rapidly replaced weighted lines and rods. As its use increased, more information about the shape of the deep ocean floor was gathered. Echo-sounders provided a two-dimensional profile of the seafloor by plotting a series of closely spaced measurements of the water depth under the moving ship (Figure 2.1). By compiling profiles and the depths from a number of sources, the shape of the seafloor was revealed despite the fact that very little of it was actually seen directly by human eyes.

There are still many portions of the oceans that have not been systematically mapped, and there are still fundamental deficiencies in how well such mapping can be accomplished. These deficiencies are

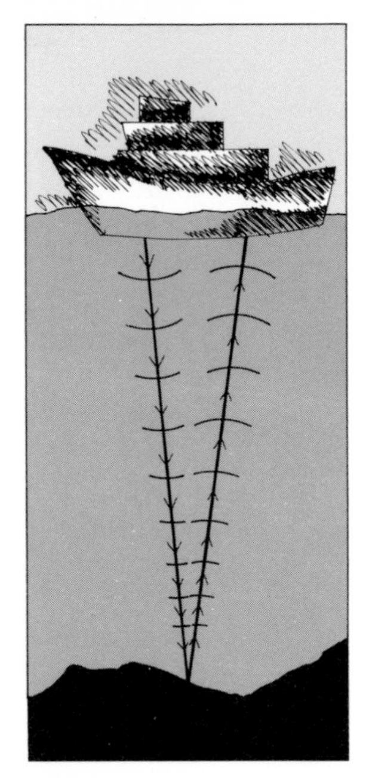

Ship taking depth sounding

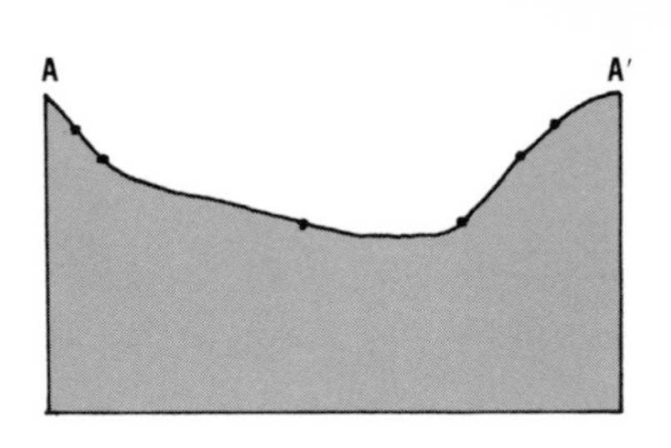

Individual bathymetric profile and series of profiles from which a map will be prepared.

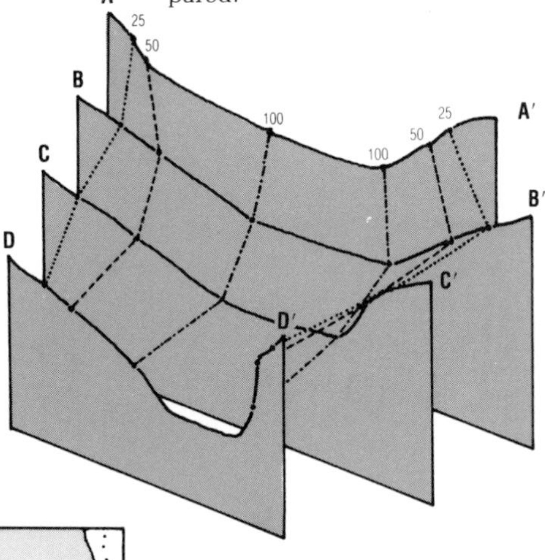

Bathymetric map showing depth contours in the Whidbey Basin between Camano Head and Sandy Point on Whidbey Island.

Figure 2.1 Compiling a profile of the seafloor using echo sounding. Top left: Individual soundings are obtained by measuring time required for sound to travel downward from the ship, reflect off the seafloor, and return to the ship. Depth is computed as half the two-way travel time times the speed of sound in seawater. Top right: A sequence of individual soundings can be plotted to make a two-dimensional, bathymetric profile showing continuous depth along the ship's trackline. Middle right: Combining several individual profiles provides basic data which is an initial three-dimensional indication of the seafloor shape. Bottom right: A map is prepared from the assembled individual profiles.

related both to navigational problems and to the echo-sounders themselves. Advances in satellite and other electronic navigation systems resolved many of the navigational problems, but obtaining precise echo-sounding measurements is still constrained by characteristics of sound waves travelling through seawater.

Exact determination of depth requires precise calculation of the speed of sound through seawater, which—although fairly accurately known—varies from place to place depending on the properties of the water. Perhaps more important is that echo-sounding is based on the shortest travel time from surface to seafloor and back to the surface. Sound waves can be focussed into relatively narrow beams and higher frequencies can be focussed more tightly than lower ones. At the same time, the transparency of seawater to sound waves is such that lower frequencies travel farther than higher ones. For deeper water this leads to a compromise in echo-sounder design. Higher frequencies and narrow beams may not be useful since attenuation of sound in deeper water may be great enough to reduce the intensity of the returning signal to a point where it is basically undetectable. Lower frequencies are less subject to attenuation but are also less focussed; they actually reflect off an area of the seafloor rather than a point. The initial return of the reflected sound can thus be a return from a point that is not directly under the ship. This limits the size of seafloor shapes that can be identified by echo-sounding—large-scale features can be determined but small-scale features, particularly in very rough areas, are much more difficult to resolve. Because it is difficult to measure water depths and define three-dimensional shapes, contemporary knowledge of the shape of the seafloor is still about 50 years or more behind our knowledge of the shape of the land surface.

Any discussion about how Puget Sound was formed involves many sources of information, principally information collected by geologists in pursuit of their broader interest in how the whole earth was formed and has evolved. Understanding differences between continents and oceans is important to our examination of how Puget Sound was formed. Some of the continent-ocean distinctions are relatively easily observed using little more than a contemporary world atlas. The most striking observation is that the earth's surface is predominantly covered by water. Furthermore, if the ocean could be removed and the earth's shape observed directly, one would be struck by two different levels: approximately 30 percent of the total surface stands at an elevation between four and five kilometers (2.5 to 3 miles) higher than the remaining area. Most of the raised portion is accumulated in a few rather large chunks, which are recognizable as continents. The continents have extreme elevations above sea level of greater than eight kilometers (4.8 miles), but average somewhat less than one kilometer

(0.84 kilometer; about half a mile). The ocean floor reaches depths below sea level in excess of ten kilometers (six miles), but has an average depth of somewhat less than four kilometers (2.5 miles).

Superimposed on this primary global scale relief are prominent features such as the mountain systems of western North America. Relief on this scale was reasonably well-defined on land by early in the twentieth century, and systematic patterns were identified that covered whole continents. Much less was known about comparable relief features of the ocean basins. Although some diversity in submarine topography was recognized early in the century, it was not until the 1950s that systematic patterns of secondary relief features in the oceanic basins were identified (Figure 2.2).

Several secondary features of submarine relief are important to understanding the development of Puget Sound. Submarine mountain ranges comparable in scale to the major continental mountains were identified relatively early. Initially, they were termed *mid-ocean* mountains because of the location of the Mid-Atlantic Ridge, which was the first such range to be described. One of the more startling dis-

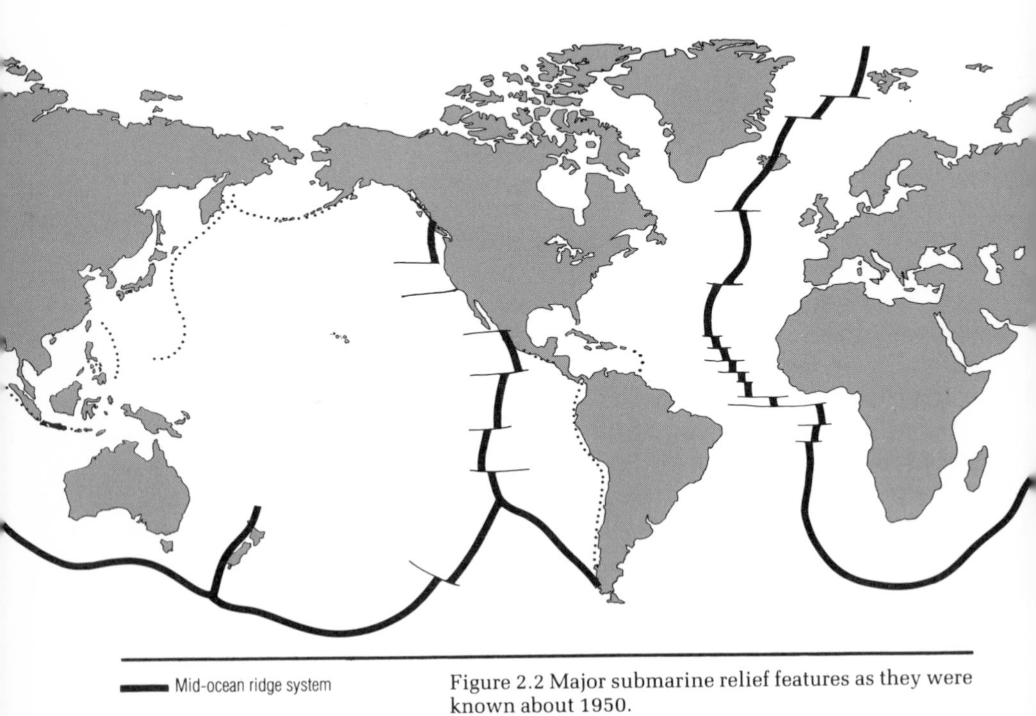

Figure 2.2 Major submarine relief features as they were known about 1950.

coveries during the mid-century marine mapping was identification of a global continuity of mid-ocean ridge systems.

Another feature of secondary submarine relief, the oceanic trench, is uniquely marine and quite unlike any feature found on land. The presence of trenches was known early in the twentieth century but systematic mapping of oceanic regions indicated several important characteristics: oceanic trenches contain the deepest depths found anywhere in the oceans, tend to be linear to arcuate in form, and are commonly associated with the edges of the deeper ocean basins.

Another aspect of secondary submarine relief proved to be one of the most perplexing features of ocean basins, irregular topography composed of long, remarkably straight, narrow bands that separate areas of relatively flat seafloor and also separate regions with slightly different average depths. These, along with trenches, are quite unlike any feature observed on land, and are termed *fracture zones.*

In the following discussion of the development of Puget Sound, it may be helpful to remember that the Sound is most directly the result of processes taking place on the margin of a continental portion of the earth's crust. Many of the larger scale processes are closely associated with processes that maintain the primary distinctions between continent and ocean. Others have either resulted in or can be best described in terms of secondary relief features. Still others are related to even smaller scale relief features, which are superimposed on the primary and secondary relief.

Description of topography, relief, and shape of the earth's surface is one of the oldest of man's descriptive sciences. As descriptive data accumulated, scientists began to categorize it and finally to ask why and how these landforms develop. Systematic exploration of the earth's surface to search for answers to these questions was one of the initial areas of study that has become the science of geology.

It is impossible to provide a brief summary to bridge the logical gap between observation of surface landforms, such as Puget Sound, and the description of the geologic development and evolution of these features. This is the corpus of geological science and is a summation of information from many diverse sources. At best, some perspective may be provided in the following pages so that the more specific description of why and how Puget Sound was formed is not presented without some context.

CHAPTER THREE

The Search for Reasons Why

The shape of natural features on the earth's surface may be described in purely morphological terms—location, size, height, elongation, and lineation—which, although they serve many practical purposes, do not satisfy human curiosity about how these features were formed. One of the geologist's most basic tasks is to understand how the crust of the earth has evolved by relating the observed end products to the geological processes that caused them (Figure 3.1).

When considering the origins of natural landforms, such as Puget Sound, it is helpful to recognize that processes that shape the earth's surface have very different scales of time and space. Some of the processes that shape the landscape are of global extent, are driven by forces originating within the earth, and result in global- and continental-scale modifications of the earth's outer crust. On a relatively short time-scale, some of the manifestations of these processes are earthquakes and volcanism; over longer time they give rise to crustal folding and faulting that result in building of mountains, such as the Cascades and Olympics, and even cause major changes in the basic form and spatial relationships of the continents and the oceans.

In contrast with global-scale processes are processes that, although ubiquitous, act in a much more localized manner. These processes derive their energy from outside the body of the earth and they act only at or very close to the earth's outer surface. Their principal source of energy is the sun. The earth's surface receives radiation from the sun, much of which is transformed into heat. Variations in heat and subsequent effects on winds and precipitation give rise to weathering of rocks, formation of soils, erosion, and transportation and deposition of sediments at the surface. All of these small-scale surficial processes modify the relief and shape of the landscape. They cause short time-scale phenomena such as floods, landslides, and dust storms, and cause cumulative impacts by, for example, modifying the course of a river valley or slowly filling a bay with river-introduced sediment.

Petrology

Studies of rocks (petrology) and their component minerals (mineralogy) traditionally provide the most fundamental component of

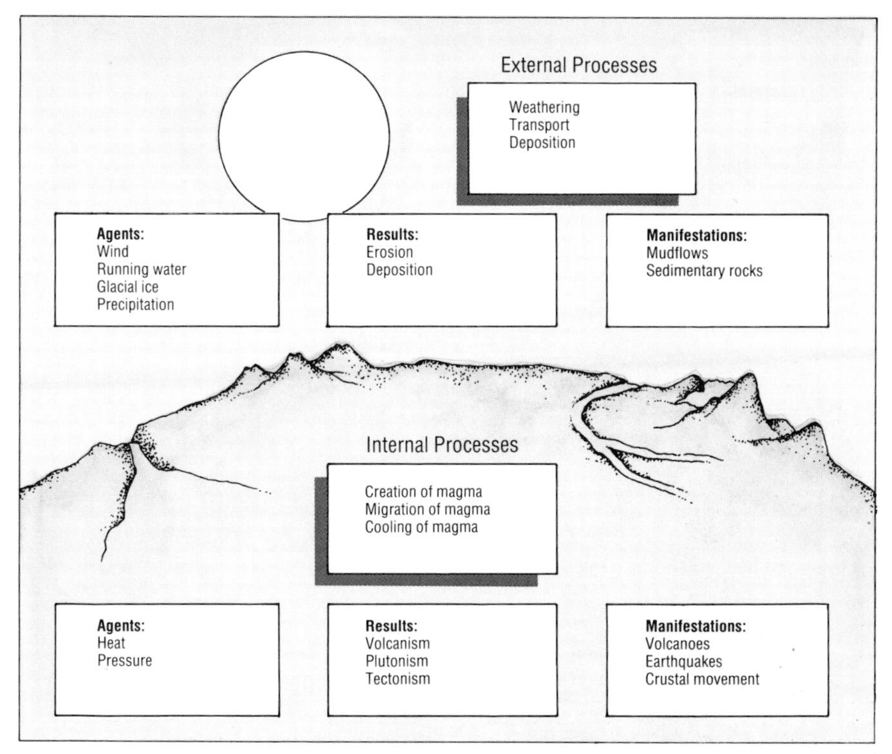

Figure 3.1 The energy driving the major geological processes is derived principally from heat, which is radiated from the sun (externally) or from geothermal heat (internally).

the geologist's toolbox. Using information about composition, spatial arrangement, and type of rocks and minerals found at and near the earth's surface, geologists have been able to answer many questions about the development and evolution of the earth's crust.

Rocks may be typed by any number of classification schemes and for any number of reasons. One of the most basic classifications used by geologists is one that implies some degree of understanding about conditions and active processes at the time and place the rock was originally formed. The principal subdivisions in this classification identify three basic rock types: igneous, sedimentary, and metamorphic.

Igneous rocks result from solidification of molten rock material as it cools. One primary element in the subdivision of igneous rock types is grain size. The cooling of molten rock material (magma) can take place at varying depths within the earth, and the rate of cooling tends to be slower at deeper depths than near or at the surface. Slower cooling permits growth of larger crystals with a resultant larger grain size in the rock. Intrusive rocks, which cool deep within the earth, display larger grain size than extrusive rocks, which have cooled more rapidly at or very close to the earth's surface.

In addition to grain size, a second primary element in subdividing igneous rocks is the chemical or mineral composition. Igneous rocks consist principally of oxides of silicon and aluminum but have varying amounts of oxides of iron, magnesium, calcium, potassium, and other elements. These latter oxides (principally iron and magnesium) form dark-colored minerals in contrast to the light-colored minerals that have higher silicon and aluminum content. As a result, silicic rocks are lighter colored, mafic (iron-magnesium rich) rocks are darker colored. Using the two basic characteristics of grain size (texture) and composition (mineralogy) a useful classification of igneous rocks has been developed (Table 3.1). With this classification, the geologist is able to infer the composition of the original molten magma as well as how and where the magma solidified.

Sedimentary rocks are the result of accumulation of rock fragments and dissolved material derived from older rocks that have been exposed at the earth's surface. The most fundamental subdivision of sedi-

Table 3.1 Major rock types

IGNEOUS ROCK

▲ INCREASING SILICA / INCREASING Fe-Mg OXIDES ▼	Coarser grained	Finer grained	Glassy	LIGHTER COLOR / DARKER COLOR
	Granite	Rhyolite	Pumice	
	Diorite	Andesite		
	Gabbro	Basalt	Obsidian	

METAMORPHIC ROCK

	Foliated	Nonfoliated
FINE GRAINED	Slate (dull, slatey cleavage)	Hornfels
	Phyllite (shiny, few visible grains) Schist (shiny, visible grains)	Greenstone (metamorphosed andesite/basalt, green color)
COARSE GRAINED	Gneiss (light colored) Amphibolite (dark colored)	Marble (recrystallized limestone) Quartzite (recrystallized chert, sandstone)

SEDIMENTARY ROCK

Nonclastic sediments (chemical)	
Classification based primarily on chemical composition	
Limestone	Calcium carbonate
Dolomite	Calcium magnesium carbonate
Chert	Silica
Salt	Sodium chloride
Gypsum	Calcium sulfate

Clastic sediments (detrital)	
Principally inorganic fragments (based on size of fragments)	
Conglomerate	>2 mm
Sandstone	1/16 mm to 2 mm
Shale	<1/16 mm
Principally organic fragments (classification based on composition)	
Diatomite	Silicious organisms
Marl	Calcareous organisms
Loam, peat	Plant debris

mentary rocks is based on whether the rock is an accumulation of fragmentary particles (a clastic sediment) or the result of chemical processes that precipitated or crystallized material that had been in solution (a chemical, or nonclastic sediment). For the clastic sediment, a principal factor in further subdivision is texture, which is basically particle size but can include additional refinements such as size range and distribution or constituent particle composition. For chemical sediment, the principal factor in further subdivision is the chemical or mineralogical composition of the aggregate. Using this classification, the geologist is able to infer the types of preexisting source rocks, and something about conditions that existed at the time and place the component particles were deposited and the sedimentary rock was formed.

Metamorphic rocks are formed when other rocks are subjected to changes in temperature, pressure, and chemical environment. The classification of metamorphic rocks considers two primary factors, foliation and grain size. Foliation is a layering or lamination in the metamorphic rock, which is directly related to the stresses that changed the preexisting rock. It results from the response of the material composing the original rock to the higher temperatures and pressures involved in metamorphism. Not all material responds in a similar manner or degree and one subdivision of metamorphic rocks is based on the degree of foliation. The grain size of metamorphic rocks varies depending upon the composition of the original rock as well as the intensity of the stressing processes to which it has been subjected. Identification of metamorphic rock type provides the geologist with some insight into the kind of stress involved (temperature, pressure, chemical changes), the intensity of the stress, and (frequently) the directional aspects of the forces that have deformed the earth's crust.

The reason for the distinctions between continents and oceans was not immediately obvious to early geologists. But over time, evidence accumulated indicating that there are basic differences between continents and oceans that are the cause of observed first-order relief features and not the result. Rocks have been sampled and collected from all over the surface of the earth. Although most of these have been collected from continental areas, a lesser number have been collected from oceanic islands, sampled by drilling cores of the ocean floor, and even dredged from the floor of the deep ocean. Mineralogical and chemical analyses of these samples indicate that there are systematic chemical differences between continental and oceanic rocks. The rocks from continental areas tend to contain a larger proportion of light-colored minerals, to be richer in silica, and to be frequently coarser grained. In aggregate, oceanic rocks are darker colored and contain higher amounts of iron and magnesium. This basic difference in composition also results in different average densities; continental rocks have an average density of 2.8 grams per cubic centimeter and are slightly less dense than oceanic rocks (3.3 grams per cubic centimeter).

Regardless of which hypothesis of the earth's origin is used, it is generally agreed that at an early stage the interior and possibly all of the planet was hot enough to be molten. As the molten mass cooled, heavier materials separated and sank while the lighter materials ascended. As a result of this separation (called *differentiation*) heavier material such as iron accumulated near the center of the earth, and lighter material such as silica accumulated in an outer layer or crust. Differentiation was a major event; without it, evolution of the earth's surface would have been different.

The light materials that make up the crust do not form a homogenous outer shell; without the contrasts between continental and oceanic rock types described above there would be no primary relief. The basic explanation for this was proposed as early as 1865 to explain observed anomalies in the earth's gravity field. The lighter density rigid crust of the earth was considered to "float" atop denser less rigid materials deeper within the earth. These deeper, less rigid materials are not fluid in the normal sense but behave as fluids because of high temperatures and pressures at the depths in which they occur. Given the concept of floating, it follows that portions of the crust with different densities and thicknesses will float at different heights, thereby displacing equal masses of the supporting medium. This concept, termed *isostasy*, was refined and modified as additional information about the earth was collected; but the basic principal remains and explains how the contrasting composition of continental and oceanic crustal rocks causes primary relief.

Geologic Age and Time Scales

Today, geologists use time units of millions and hundred-millions of years in the same casual manner that economists use comparable numbers of dollars. Quantities of these magnitudes are not actually comprehended by anyone, but the units are accepted in common use. Contemporary techniques for determining the absolute age of crustal rocks depend upon decay rates of radioactive minerals and isotopes occurring naturally in rocks. Measuring accumulation of decay products that have formed at known rates permits an estimate of the age of the rock, the time that has passed since the rock and the incorporated minerals were originally formed. These techniques are not very old, but they confirm the fact that geologic time scales are *very* long.

To decipher past history, it is necessary to place events (in this case, identifiable results of ongoing geologic processes) in chronological sequence and then to determine an absolute age for the individual events. When geologists first began to study the earth, their efforts were localized and directed more at determining relative ages than absolute ages. Determination of sequence began with recognition of the principal of *superposition*. Simply stated, this principal says that if something (e.g., a mudflow, volcanic lava, sediment accumulating in a lake) is laid down on an existing surface, that which is laid down is younger than the surface it covers. This led to development of additional principals for determining older-younger relationships and, in time, a body of information that established relative ages of rocks and other geological features. The systematic accumulation of relative ages over large surface areas permitted synthesis of the history of the earth's crust—how it was formed, how it has been changing.

An accurate awareness of the absolute age of the earth and the many events involved in its history was relatively slow in developing. Before the middle of the eighteenth century, a general concensus placed the absolute age of the earth as several thousands of years. In order to squeeze all of the events recorded in the geological record into that period of time, geologic processes had to work at rates much faster than actually observed or to be intermittently impacted by worldwide catastrophes. But there was no rational reason to expect the rates of geologic processes to have suddenly slowed down just about the time geologists became interested in them. Further, although catastrophes did occur, they were relatively local in extent and did not happen with anything resembling the frequency required to squeeze all observed historical geological events into a few thousand years. Consequently, by the late eighteenth century, it became generally accepted that measured rates of geologic process were representative of what they had been throughout time. This led to acceptance of the concept that the earth's age was vastly greater than previously believed—perhaps hundreds of thousands of years. By the late nineteenth century, estimates of the age of the sun and the solar system placed the possible age of the earth at as much as tens of millions of years, and by the early twentieth century, radioactive dating techniques raised the total to hundreds of millions and even billions of years.

Absolute age determinations of rocks collected from the ocean basins were puzzling. There was an absence of any really old rocks in any of the oceans. The oldest rocks in the Pacific Ocean, for example, were only 180 million years old, a sharp contrast to continental rocks, some of which were several hundred million years old. In addition to being younger, there were systematic sequences of increasing age in some oceanic areas. This was most obvious in the northern and equatorial Pacific where the age of rocks generally increased from east to west.

Geophysics

In contrast with observable information on the shape of the earth's surface or the composition of surface rocks, much information about the earth has had to be obtained indirectly. Although the shape of the surface can be measured and samples of surface rocks collected and analyzed, many of the forces that shape the earth's crust result from the larger scale processes originating within the body of the earth. Although these forces and processes cannot be observed or measured directly, many of them can be inferred from measurements taken at the surface. This type of investigation is part of geology and is the special field of the geophysicist who uses a wide variety of indirect measurements to conduct such investigations.

During the 1950s there was a significant increase in systematic measurement of geophysical phenomena all over the world, particularly in the marine area. For the first time, the earth's gravity field, magnetic field, and heat flux were measured, and rock samples were collected to investigate broad portions of ocean floor. Prior to this time, geological thinking about the origin of crustal features had been based almost entirely on observations made on land. Expanding geological investigations from continents into the oceans—in order to include the remaining two-thirds of the earth's surface—resulted in new insights into earth processes. Key findings by marine geologists and geophysicists provided impetus for new concepts of how the earth's crust evolved.

Seismology

One of the oldest branches of geophysics is seismology, the study of earthquakes. Earthquakes are one of the most catastrophic responses of the earth's crust to ongoing stresses developed by processes originating within the body of the earth. Because earthquakes can have strong impacts on society they have been reported for many years. The study of earthquake phenomena, seismology, began relatively early and has evolved into one of the principal tools for investigating the internal structure of the earth and its crustal rocks. From careful examination of collected records, seismologists contributed to understanding of the earth's internal structure and insight into how the earth's crust has been formed and modified.

Earthquakes originate at varying depths within the earth. When lithospheric rocks that have been subjected to deformational stress fail suddenly (i.e., break) they rebound rapidly toward their unstressed positions, generating shock waves at the site of the break (called the focus or hypocenter). These shock waves move outward from the focus both through the body of the earth and along its surface. The paths that shock waves travel are determined by the speed of the waves, which is in turn a function of the elastic properties of the earth material through which they travel. Characteristics of amplitude, wave motion, and time of arrival are recorded by special instruments called seismometers. Seismologists have compiled much information about the interior of the earth by systematic examination of seismometer records (seismographs). By comparing arrival times of earthquake waves received at seismometers in different locations, the epicenter can be determined. One of the early contributions of seismology was the location of the epicenter (a point on the earth's surface directly above the focus of the shock) and depth of focus of an earthquake (Figure 3.2).

Ongoing improvement in instrumentation and data analysis methods have refined basic seismological procedures and, in addition to epicenter and depth of focus, seismologists now can be reasonably precise about the characteristics of the crustal rock movements that originated the quake. Examination of systematic compilations of this kind of data has provided insight into a fundamental characteristic of the crust—most earthquakes tend to be located in long narrow bands which, although interconnected, surround relatively large regions that are basically aseismic (i.e., without earthquakes). Of particular interest was the observed strong correlation between the seismic belts and the secondary relief features of the ocean basins (submarine mountains, trenches, fracture zones, which were cited in Chapter 2).

Since seismic shock waves move through the earth at different speeds and along various paths, much has been learned about the earth's interior by careful study of seismic records. By the middle of the twentieth century there was almost universal agreement about the gross features of the earth's interior. A basic layering was identified, based on study of shock wave transmissions. A central core and a surrounding mantle occupy almost all of the body of the earth, and an outer shell containing about one percent of the total earth volume and only tens of kilometers in thickness forms the thin crust that encloses it. In addition

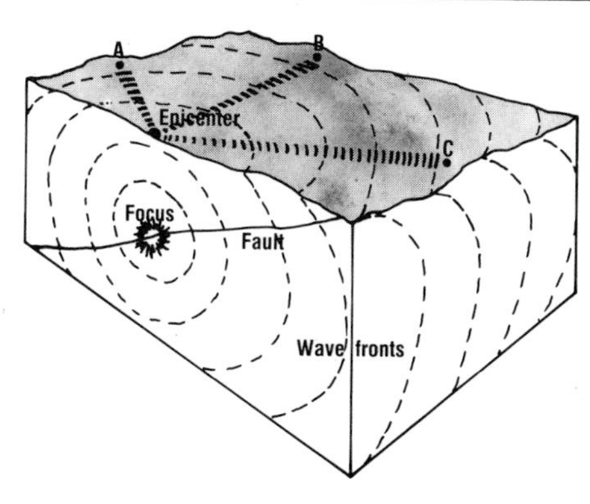

Figure 3.2 The source of energy released during earthquakes is the focus from which seismic wave fronts propagate outward. At seismograph stations (A, B, C) time of arrival and seismic wave velocity are recorded and used to locate the position of the epicenter.

to contrasting rock types, other differences between continental platforms and oceanic basins have been identified seismically. Perhaps the most important discovery was recognition that the crust was generally much thicker under the continents than under the oceans.

On a smaller scale, using shock waves initiated by small explosions or other active sound sources, seismologists have employed similar analytical techniques to define the details of near-surface structures.

Geomagnetism

The earth has a magnetic field (the geomagnetic field) which can be approximated by a strong bar magnet located within the body of the earth and aligned almost parallel to its rotational axis. Shape and intensity of the geomagnetic field can be measured at the earth's surface by various instruments, the most familiar of which is the compass. Because of the relationship of the positions of the geomagnetic and geographic poles of the earth, the compass has been used for several centuries to indicate direction on the surface of the earth. A similar instrument, called a dip-needle, swings in a vertical plane and indicates the vertical component of the geomagnetic field. More sophisticated instrumentation, called magnetometers, can provide additional data on the intensity of the geomagnetic field at any measurement station.

Measured field intensity varies considerably from place to place because—in addition to the geomagnetic field—local magnetic field intensity is affected by magnetization of crustal rocks. This magnetization is a basic property of rocks and is related to their thermal history. When the temperature of magnetizable material is higher than about 500° Celsius (the Curie Point), magnetic components are free to align themselves with the geomagnetic field in much the same way as bar magnets in a compass or dip-needle. When the temperature cools below the Curie Point, alignment of the magnetic components is frozen, they are no longer free to move, and their alignment remains fixed as long as the temperature does not once again rise above the Curie Point. This type of magnetization (thermoremanent magnetization) is a property of many rocks found at and near the surface of the earth. Because thermoremanent magnetization is determined by the characteristics of the geomagnetic field that existed at the time and place a rock cooled from above the Curie Point, it has been extremely useful in deciphering aspects of the history of the earth's crust.

Systematic investigations of rock magnetism from many places around the earth has indicated that many rocks possess a thermoremanent magnetization that is not aligned with today's geomagnetic field. This could be explained two ways: either the rock moved along the surface of the earth after it was magnetized, or the geomagnetic field changed after the rock cooled below the Curie Point. Actually, contemporary thinking considers that both processes have occurred and sorting them out has provided some of the strongest confirmation for current hypotheses on the structural evolution of the earth.

Widespread use of magnetometers in the oceans started in the 1950s. Magnetometers measure local magnetic field strength, a summation of the large-scale geomagnetic field and magnetization of local crustal rocks. Using a technique that removes the contribution of the smoother geomagnetic field, the localized effect of the crustal rocks can be observed as a *magnetic anomaly*. Patterns formed by the marine magnetic anomalies were initially quite unexpected since no similar conditions were observed on land. Characteristically, oceanic magnetic anomalies are linear, occur in roughly parallel bands that display remarkable continuity over long distances, and are oriented approximately parallel to the mid-ocean ridges. Since these lineations reflect the patterns of magnetization of oceanic crustal rocks they provide one of the final keys to refining a generic model that explains the structural evolution of the earth's crust.

Crustal Evolution

Prior to the geological information explosion that occurred after 1950, theories about the formation and evolution of major structures of the earth's crust were relatively diverse and, it is fair to say, frequently provincial. This was understandable since most of the data upon which earlier hypotheses had been based were observations from the continents, and even these were biased by European and North American conditions. As additional observations became available, particularly from oceanic areas, there was an increased awareness of additional contrasts between oceanic and continental crust. Perhaps most important was the determination that some crustal processes (and their resultant submarine topographic features) were uniquely oceanic. Examination of these processes in combination with available information provided the initial opportunity to develop hypotheses about evolution of all of the earth's crust, instead of only the continental portion.

Seafloor Spreading

By the early 1960s many bits of information had been collected—many of them during the previous decade—that combined to give impetus to a concept of lateral mobility of the oceanic crust. Termed *seafloor spreading*, this concept proposed that mid-ocean ridges were sources of new oceanic crustal rock emplaced along the ridges by submarine volcanic activity. As new crust continued to be added along the ridge, the existing oceanic crust was displaced and moved away from its axis. Over time, as this process continued, the seafloor spread lat-

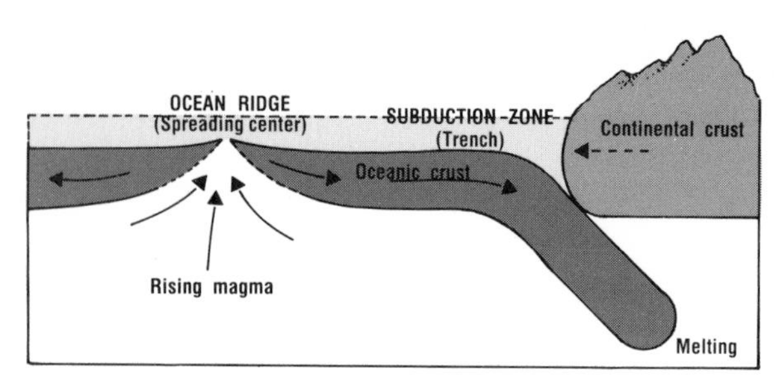

Figure 3.3 As new crust is formed along an oceanic ridge, older crust moves laterally away from the spreading center. At its older (outer) edge, the dense and heavy oceanic crust moves under the less dense, continental crust. As it is subducted the temperature increases and it ultimately melts and is mixed with underlying rock.

erally away from the ridge (Figure 3.3).

Clearly, if new crust was being formed, there would have to be some way to destroy old crust or the earth would continually grow larger. According to the seafloor spreading concept, oceanic crust is destroyed at *subduction* zones corresponding to oceanic trenches. Along these zones old crust moves downward toward the deeper portion of the earth, melts, and becomes incorporated into the underlying mantle. Although this hypothesis did not directly address problems involving oceanic and continental crustal interactions and contrasts, it did account for phenomena in ocean basins, such as mid-ocean ridges and trenches, that had been defined during the prior decade. Three contributions of the concept were identification of mid-ocean ridges as sources of new oceanic crust, explanation of lateral movement of oceanic crust that was progressively older away from its source, and destruction of old oceanic crust at subduction zones.

Plate Tectonics

With additional refinements to explain operation on the spherical shape of the earth's surface, the seafloor spreading hypothesis evolved into the "global plate tectonics" hypothesis. This hypothesis received general acceptance during the 1970s, and provided a framework for discussing the evolution of western North America and the formation of Puget Sound. Its premise is that the outermost portion of the earth is enclosed by a shell of material termed the *lithosphere*. The lithosphere is cool and brittle (rigid) when compared with the underlying *asthenosphere*. The outer surface of the lithosphere may be either oceanic or continental crust, and the lithospheric shell itself is broken into a number of separate pieces which are called plates. These plates are rigid and interact with adjoining plates.

A key element in this concept is that there is little happening to the plates themselves *except* along the plate boundaries. Since plate boundary interactions give rise to seismic activity, principally earthquakes and volcanism, the boundaries between plates can be identified on a map showing the worldwide distribution of seismic activity, volcanism, or seismic belts (Figure 3.4). Interaction of one lithospheric plate with another gives rise to plate boundary phenomena that differ depending on the relative movement of opposing plates. This movement can be of three types: divergent, shear, or convergent (Figure 3.5).

Divergent boundaries occur when new oceanic crust is formed and

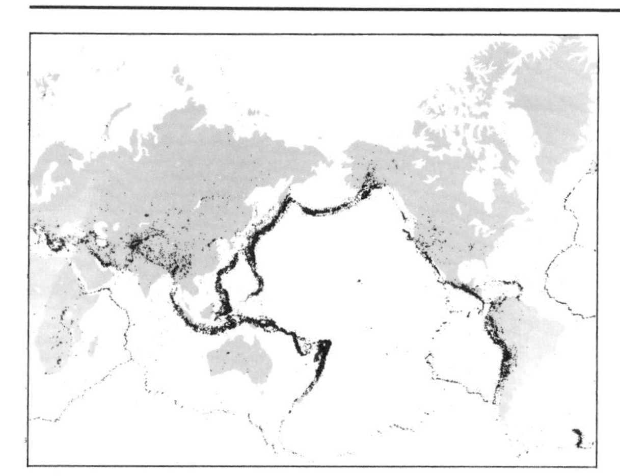

Figure 3.4 Global crustal plates and distribution of earthquake activity. Below: The principal crustal plates are rigid and have very little seismic activity in their interior portions. Left: Global distribution of seismic activity correlates with the boundaries of individual plates (courtesy National Oceanic and Atmospheric Administration.)

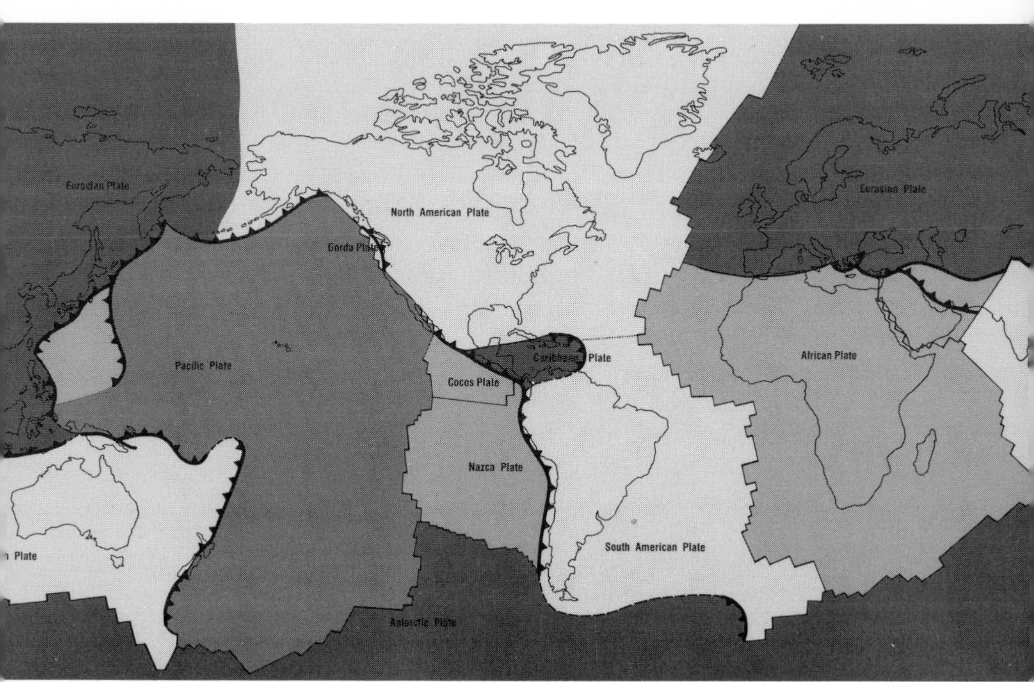

Figure 3.5 Types of crustal plate boundaries: divergent, convergent, and shear. *Divergent boundaries* are associated with shallow focus, low energy earthquakes and volcanism generated by the formation of new crustal material. *Convergent boundaries* have earthquakes, ranging from shallow to deep focus that are associated with shearing movements between the subducting slab and the overlying crust. Volcanism can be massive and magma can cool well below the surface. Surface volcanism is associated with the "magmatic arc," a position at the earth's surface directly above the point where the temperature of the descending slab gets high enough to melt. *Shear boundaries* are where seismic activity can be of very high energy, but the focus is generally shallow (confined to the thickness of the shearing plates). Characteristically there is no volcanism along shear boundaries.

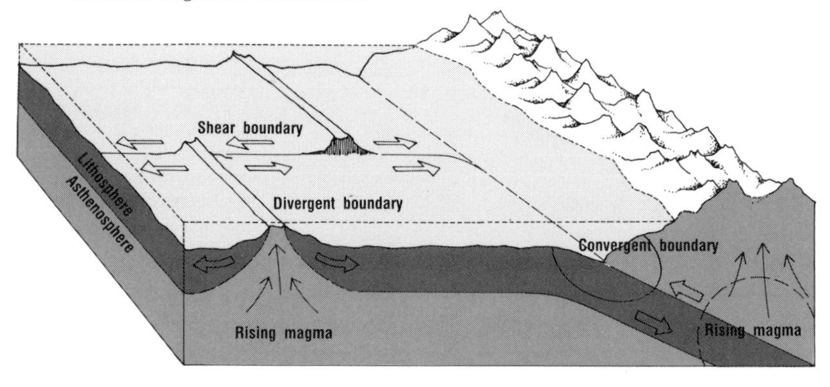

the adjoining plates diverge from spreading centers associated with mid-oceanic ridges. These boundaries are characterized by volcanism, which gives rise to new crust, and by seismic activity that tends to be of low level and shallow focus because it is confined to the thin new oceanic crust.

Shear boundaries occur where adjoining plates move laterally along the boundary, and frequently correspond to oceanic fracture zones. Crustal material is neither formed nor consumed at these boundaries. Seismic activity at shear boundaries does not involve volcanism and is marked by large, shallow-focus earthquakes that result from lateral displacement.

Convergent boundaries occur where plates move together. They are commonly associated with deep-ocean trenches, high level of seismic activity (both shallow and deep), and volcanism. Convergent boundary processes are of most interest to a discussion of Puget Sound because most of the crustal development of western Washington involves the interaction of oceanic and continental crust along a persistent convergent plate boundary.

Convergent plate boundaries differ in detail depending upon whether continental or oceanic crust is involved. Three types of plate convergences may occur: oceanic crust may converge with oceanic crust, oceanic crust may converge with continental crust, or continental crust may converge with continental crust. Each of these three types has been active at some time during the evolution of western North America and each has left its imprint on the structure of the Puget Sound region.

When two oceanic plates converge, one or the other is underthrust, and magma forms when the downthrust crust reaches a temperature high enough to cause melting. This magma, primarily basaltic, rises and is subsequently extruded to form a characteristic volcanic island arc. As the process continues, the arc grows in size and erosional debris from the arc is deposited on either side of the arc, the aggregate forming an *arc terrane*. In late stages of development the arc terrane can be deformed and uplifted to form new continental crust. This class of convergent boundary processes gave rise to the arc terrane and continental crust that played an important role in the development of Vancouver Island and the Coastal Mountains.

When convergence involves interaction of continental and oceanic crust, the heavier oceanic crust is normally overridden by the lighter continental crust. Structural deformation on the continental crust frequently involves development of mountains and associated volcanic activity that are different from those resulting from convergence of oceanic and oceanic crust. This class of convergent process has played an important role during much of the development of western North America and is continuing in the contemporary development of Washington and Oregon.

When continental and continental crustal material converge the result is characteristically the accretion of one to the other, the "clogging" of the subduction zone and a consequent realignment of the interplate boundary. This class of convergent process, involving continental-type aggregates carried along with the oceanic plate to the subduction zone, was important to the development of western North America and specifically the Coastal Mountains and Vancouver Island.

CHAPTER FOUR

Puget Sound—The Past

Although the face of the earth has been changing continually through geological time, the location and form of Puget Sound is the product of only the last few hundred millions of years. Because geological processes are ongoing, contemporary Puget Sound is a transient feature of the ever-changing geography of the earth.

There is a general correlation of time and space that is helpful to examining how Puget Sound was formed. An expanded "address" for Puget Sound will place it in a spatial (size) continuum of topographic features:

Puget Sound
 Puget Lowland
  Western Margin of North America
   Earth
    Solar System

On a global scale, Puget Sound is a feature of the continental portion of the earth, rather than the oceanic portion. On a continental scale, Puget Sound is a feature of the western margin of North America. On a regional scale Puget Sound is the flooded portion of the Puget Lowland.

Development of the crustal structure of western North America began more than 300 million years ago but the structural identity of the Puget Lowland was established only as recently as 20 to 40 million years ago. This very slow development of crustal structure was the result of geological processes acting over tens to hundreds of millions of years and affecting broad areas of the earth. Once the large-scale structure of the Puget Lowland was established, the detailed, small-scale surface topography was shaped. This was accomplished primarily by surficial processes, the most prominent one being glaciation during the ice age of only a few tens of thousands of years ago.

Evolution of Western North America

With even a cursory look at an atlas, a globe, or satellite photographs, one can see the American Cordillera—that belt of the earth's crust extending along the western edge of the Americas from the Aleutians to the southern tip of South America (Figure 4.1). The North American portion of the Cordillera covers about one-third of the conti-

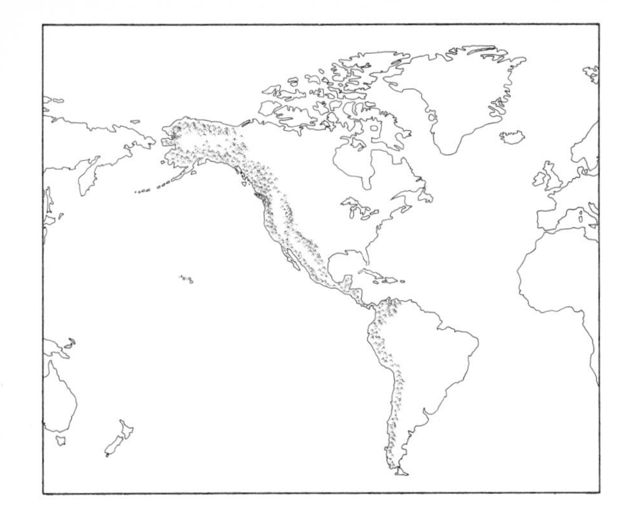

Figure 4.1 The American Cordillera.

nent from the Rocky Mountains to the Pacific Ocean. It is a complex belt of mountain ranges, intermountain basins, and plateaus that comprise the highest and most rugged topography of the Americas. The principal subdivisions of the North American Cordillera are prominent linear mountain chains to the east and west and between these ranges a region of interior mountains and plateaus. It was the long, narrow belts of mountain ranges—the Rocky Mountains, Sierra Nevadas, and Cascades—that inspired the term Cordillera from the Spanish word for "cord" or "rope."

Although there are many complex, and sometimes conflicting details known about the Cordillera, a gross generalization of rock types and existing structures can be made. Along the eastern border, there is a north-south band of primarily sedimentary rocks that were deformed structurally but were not subjected to the extreme deformation found farther west. The sediments that formed these rocks accumulated in shallow marine water, contain few coarse materials and no indications of volcanic activity, and have a sediment distribution that implies the presence of continental crust to the east during the time they were accumulating.

The western Cordillera is a strong contrast to its eastern counterpart. It is a belt of mixed rock types, contains evidence of more severe metamorphism, coarse sediment intermixed with volcanic rocks, indications of accumulation in deeper water, and volcanic rocks that imply a submarine origin. This sequence in the petrotectonic assemblages—a combination of rock types, structural relationships, and spatial or geographic arrangement—indicates growth along western North America resulting from convergent plate boundary processes over the past 300 million years. Westward growth was not accomplished in a single step, however; it resulted from an aggregate of many events along the ancient continent's western boundary.

The contrast beween the east and west sides of the Cordillera appears to have persisted throughout much of its development. For rocks older than 150 million years, relationships among different petrotectonic assemblages have been obscured by more recent structural deformation—folding, faulting, metamorphism, and batholithic intrusion—as the convergent plate boundary processes moved progressively westward. Today, the eastern portion of the continent and the Cordillera is relatively stable, but the western Cordillera is still an area of active crustal processes.

The Earliest Beginning

The structural development of the Cordillera began about 350 million years ago with an apparent change in plate boundary conditions along the western margin of ancestral North America. Prior to that time conditions were quite the opposite of what is observed today. The eastern portion of the continent was being developed by convergent plate processes, the Appalachian Mountains were being built, and the continent was moving toward an imminent collision with other continents (ancestral Europe and Africa). The western portion of North America was not a plate boundary. Continental and oceanic crust were coupled and were part of the same plate moving away from a now lost spreading center in the oceanic region to the west. In this respect, although east and west directions were reversed, the ancient western continental boundary was similar to the present boundary between eastern North America and the Atlantic Ocean.

From within the region associated with today's Puget Lowland, there are local rocks that are more than 350 million years old. They are not widespread and their age is generally indeterminate, other than that they are older than age-dated rocks that overlie them. Some of these rocks occur in the San Juan Islands, some occur in a discontinuous band along the southeast end of Vancouver Island, and some are found in the western Cascades near Marblemount on the Skagit River. They all display evidence of metamorphism, some quite severe, but provide little insight into local history other than that they are old and for many years were considered to form a "basement" for overlying rock.

About 350 million years ago there was a reorganization of the existing boundary on the oceanic plate somewhere west of the continent. This event resulted in the Antler Orogeny (named after Antler Peak in Nevada) and was the earliest decipherable event in the structural evolution of the Cordillera. It marked the first addition of new crustal material against the western margin of the continent and represented the beginning of convergent plate boundary processes that have continued to this day.

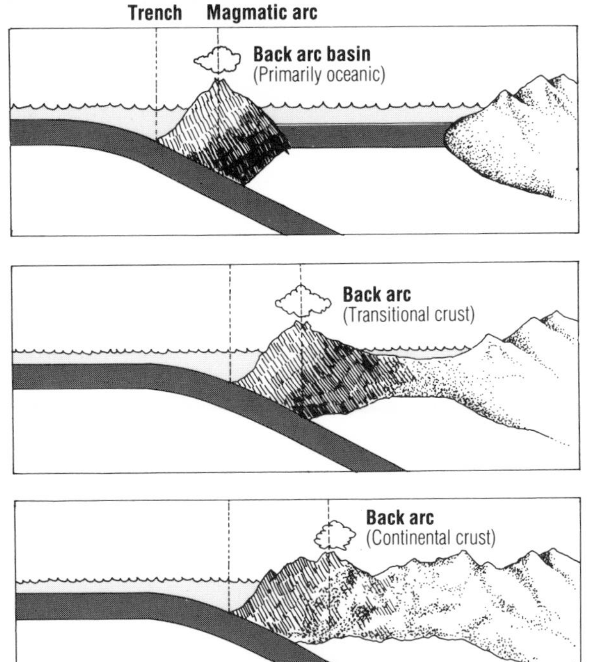

Figure 4.2 Evolution of the western North American boundary (approximately 225–65 million years before present) from Japan-type to Andean-type. Top: A Japan-type convergent boundary was characterized by an offshore magmatic arc that was separated from the continent by a marginal oceanic basin (back arc basin). Middle: During the transition, the marginal basin was modified by sedimentation from both the arc and the continent, and was shortened, deformed, and compressed as the arc was pushed up toward the continent. Bottom: When modification of the former marginal basin was completed, the arc terrane became part of the continental mass and the transition to Andean-type was completed.

The Japan-type Boundary

With the initiation of these convergent boundary processes, development of the Cordillera began. The western shoreline of the continent began its extended migration to the west. For about 150 million years the convergent boundary at the western margin of North America resembled the situation that exists today in the western Pacific. There was an offshore trench where oceanic crust was subducted, an offshore volcanic island arc, an extensive back-arc basin where detritus from the island arc (sediment and volcanic debris) accumulated, and a foreland basin that bounded the continent from which sediment was derived. These processes and features have a contemporary analogy (albeit with east-west direction reversed) in a section extending from the northwest Pacific, across the Kuril-Kamchatka Trench, the Kuril Islands, the Sea of Okhotsk, and onto the Asian mainland. Based on today's geography, a convergent boundary of this kind is called *Japan-type* (Figure 4.2).

Rocks dating from this period are found locally in the San Juan Islands, in the western Cascades, and on Vancouver Island. In aggregate they are a mix of sedimentary rocks interbedded with volcanic rocks, representing marine accumulations of a suite of rock types commonly associated with offshore volcanic islands. In themselves they provide very little insight into the structural evolution of the Cordillera; but they do imply crustal instability at their accumulation site, and the association of shallow water deposition, volcanic debris, and submarine volcanic flows is consistent with convergent plate boundary processes.

By about 200 million years ago an ancestral Rocky Mountain system existed along the western coast of North America. The coast was still far east of the present west coast and what is now the western Cordillera was still accumulating under marine conditions in an area marked by crustal instabilities associated with the converging plate processes. This initial phase in the development of the Cordillera came to an end marked by the Somona-Tahltanian Orogeny, which occurred between 250 and 180 million years ago. This was a widespread event characterized by eastward thrusting of sediments onto the continental margin (the Somona Orogeny in Nevada) and an interval of granitic intrusions (the Tahltanian Orogeny in British Columbia).

About 200 to 180 million years ago there was a worldwide reorganization of lithospheric plate movements. The plate movements of the prior several hundred millions of years had resulted in the accumulation of continental crust in a super-continent called *Pangea*, which because of the ongoing nature of crustal evolution existed for only a relatively short time. Initial rifting of this proto-continent took place in what is now Scandinavia about 250 million years ago and over millions of years eventually resulted in the separation of North America from Europe and Africa. Continents identifiable by today's geography were separated completely by 100 to 80 million years ago (Figure 4.3), a period during which there was major accretion onto western North America. The Cordillera, with the exception of the extreme western portion, took on a form that is recognizable today.

The movement of continent was related directly to lithospheric plate movement and many of the basic plate interactions are still con-

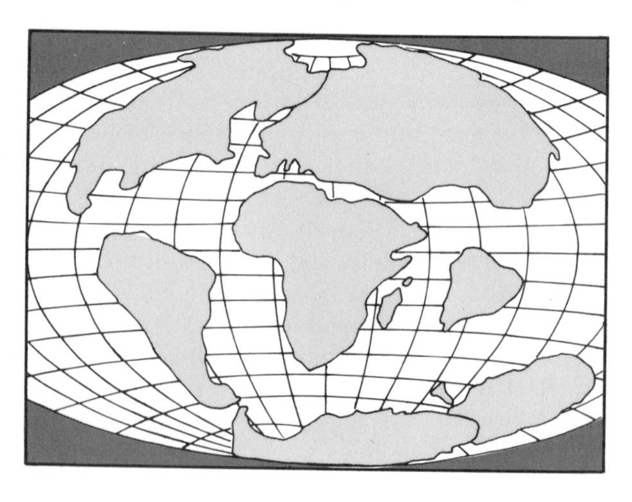

Figure 4.3 Pangea as it looked about 65 million years ago (After Deitz and Holden 1970).

tinuing; there is some late rifting in the Red Sea area and East African rift valleys and continuing crustal divergence along the Mid-Atlantic ridge. Convergent boundary processes that built the Cordillera westward from the Rockies are still continuing along part of the western boundary of North America.

The Japan-Andean Transition

At the time of the initial breakup of Pangea, the western boundary of North America was still Japan-type; however, over the next 100 to 150 million years, during the Mesozoic Era (225 to 65 million years ago), conditions along the boundary underwent a major transition so that by the end of the Mesozoic it was of the *Andean-type*. This is a convergent boundary named after the features, processes, and conditions occurring today west of Chile and the Andes Mountains. The transition from Japan to Andean-type resulted primarily from shortening and modifying the back-arc basin (Figure 4.2).

Much of the regional history for this transition period is based on analysis of the relationships between magmatic arc and subduction zone rock assemblages as well as delineation of crustal shortening in thrust belts east of the magmatic arc. At the beginning of the transition the magmatic arc of western America was nearshore at the south but still separated from the continental platform at the north by a large back-arc basin. The shortening of back-arc features began in the south and progressively shifted northward to accomplish the transition.

The Mesozoic magmatic arc was the site of some of the most massive intrusion of igneous rock on the continents (Figure 4.4). For more than 100 million years the arc was repeatedly intruded by molten rock originating deep in the convergence zone. This repeated igneous activity added up to a massive addition of new crustal material in the form of andesites and granites. The Coast Batholith of western British Columbia, the Idaho Batholith, and the Sierra Nevada Batholith all date from this period.

The magmatic arc shifted laterally throughout the Mesozoic but maintained the general shape shown in Figure 4.4. The position also shifted—sometimes closer, sometimes farther away—from the subduction zone. This shifting can be attributed to changing subduction angles of the oceanic plate influenced by the changing convergence rate and resulting in shifting of the position at which the magma melts.

Rocks formed during the Mesozoic are found in the San Juan Islands, in the Cascades, and on Vancouver Island. In general, local Mesozoic rocks are a thick sequence of marine sediments and volcanic rocks that contain evidence of increased igneous activity (volcanism and granitic intrusion) during the mid-Mesozoic. Petrotectonic assemblages imply that the rocks were formed in a region of crustal instabil-

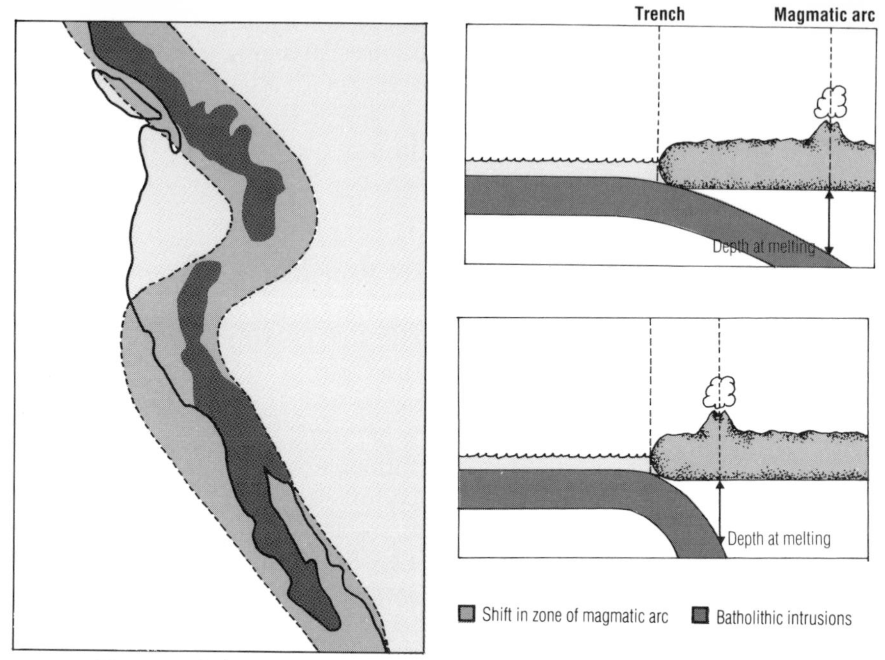

Figure 4.4 Left: During the Mesozoic (225–65 million years before present) the axis of the magmatic arc shifted laterally within the indicated zone (after Dickinson, 1976). Right: The changing position of the axis of the magmatic arc relative to the trench is related to changing of the angle of the subducting slab. This appears to be a function of changes in the rate of plate convergence.

ity, were affected by mid-Mesozoic igneous activity and were subjected to major stresses late in the Mesozoic.

The region now occupied by Puget Sound was still relatively far to the west of the continental shoreline. Conditions were oceanic through the Mesozoic, and the subduction zone and associated magmatic arc trended northwest-southeast.

One of the more puzzling aspects of local geology is the anomalous magnetic pole positions from some rocks on Vancouver Island. Magnetism measurements from basalt rocks on Vancouver Island indicate that they were formed several hundred kilometers south of comparable-age rocks to the east. These basalt rocks were formed early in the Mesozoic, and the anomalous magnetic pole positions imply that they must have moved far north before being accreted to the continent.

The concept of accretion (i.e., addition) is important to understanding the development of the western Cordillera. An important process in establishing today's crustal structure was the accretion to western North America of some large chunks of nonoceanic crust that had originated on the plate west of the subduction zone. Where oceanic crust and continental crust converge, heavier oceanic crust normally moves beneath continental material. On the other hand, if convergence brings together crustal material of comparable density, the result is more complex. When the converging oceanic crust carries with it a large chunk of lighter material, such as a piece of continental crust or an island arc, the subduction zone may become overloaded and clogged. The contact zones between converging sectors of lighter crust, termed the *suture*, frequently contain distinctive petrotectonic assemblages, and mark the approximate site of the clogged subduction zone.

Anomalous magnetic conditions in volcanic rocks on Vancouver Island can be explained by accretion of relatively large insular terranes about 190 to 160 million years ago (Figure 4.5). These originated on the oceanic plate west of the subduction zone and were carried to their present relative position by plate convergence. Following the accretion of insular terrane, the subduction zone shifted to the west.

A long period (from 155 to 95 millions years ago) of deposition in a prominent forearc basin followed, but by the close of the Mesozoic the magmatic arc shifted westward, and the shift to an Andean-type boundary was essentially completed. By the close of the Mesozoic (65 million years ago), most of what is now the western and central Cordillera was in place. The western coast of North America was bordered by high mountains capped by active volcanoes. A short distance offshore a trench marked the axis of the active subduction zone. The coast was close to the south shore of Vancouver Island, but swung well inland to the east of the current Washington and Oregon coast in a large embayment.

The Andean-California Transition

Between 60 and 40 million years ago a prominent reorganization of the global crustal plates occurred. Plate boundaries both north and south of the large embayment began to modify from convergent to shear boundaries. This was partially due to plate movement changes and subduction of portions of the oceanic spreading center that had persisted over the previous several hundred million years. These changes began a modification of the continent-ocean margin of western North America from Andean to California-type.

Although convergent motion became increasingly anomalous along the western margin of North America, it persisted off Washington and Oregon. Consequently, examination of the final local crustal evolution will still be in terms of convergent boundary processes but on a more localized scale than our earlier examination of broad development of the Cordillera.

By about 50 million years ago the spreading center for oceanic crust was still several hundred kilometers west of the continental mar-

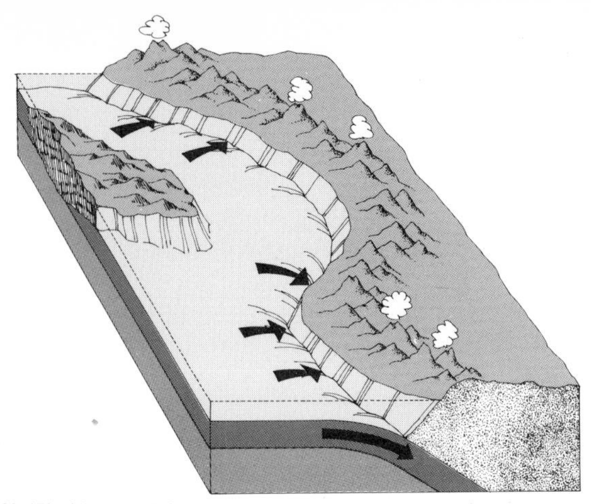

Figure 4.5 Accretion of Vancouver Island to ancestral North America.

A continental fragment or island arc terrane is carried along with the oceanic crust northeastward toward the subduction zone which during this time extends from Northern California to British Columbia.

180–170 million years before present

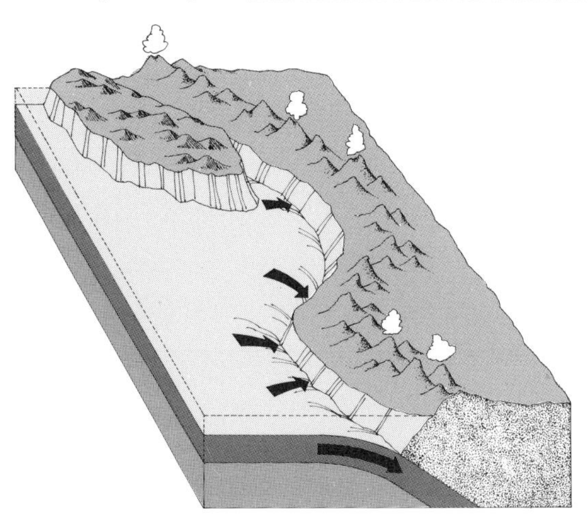

As it bumps up against the continental crust the continental fragment is too large and too light to be subducted along with the oceanic crust. The subduction zone is temporarily clogged and local subduction ceases.

160–150 million years before present

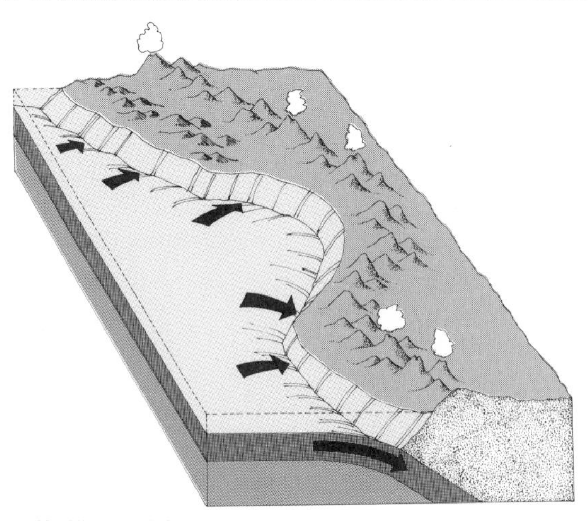

As convergent plate movement continues the continental fragment becomes accreted to the main continental mass, and the subduction zone "jumps" to a new position on the oceanic side of the accreted fragment.

150–130 million years before present

gin. The oceanic crust moving eastward from this center was feeding into a subduction zone at the convergent boundary with the North America plate. The axis of the subduction zone swung well to the east of the present continental margin into the large regional embayment. Although there is no clear evidence in contemporary surface rocks, the axis of the subduction zone may have swung to the south along a position occupied today by the Puget Lowland or along the western flank of the Cascades. The associated magmatic arc was farther east, located approximately near what is now Idaho. Between the subduction zone and the magmatic arc a nonmarine forearc basin was accumulating material that is now found in the northern Cascades. This terrestrial detritus makes up, for example, much of the several-thousand-meters-thick accumulation of rock that is exposed at some places on the western flanks and more on the eastern side of the Cascades.

About 40 to 30 million years ago, a mass of marine basalts and sediment, pillow lavas, and pelagic limestones that had originally accumulated on the oceanic plate moved eastward into the region of the subduction zone (Figure 4.6a). In a manner similar to the process that earlier had formed Vancouver Island, these materials were accreted to the continent, forming what would become the Coast Range extending from northern California to central Washington. It also resulted in a westward shift of the subduction zone, removing most of the concavity from the trench axis that had been associated with the large regional embayment. A small remnant of the earlier indentation of the continental margin remained between Vancouver Island and the central Washington coast. The relocation of the axis of the subduction zone gave rise to a westward shift of the magmatic arc to its present position along the trend of the Cascade Mountains. The repositioned magmatic arc established the locus for volcanism in the Cascades that has been active intermittently up to the present day.

The westward shift in the subduction zone also gave rise to a prominent forearc basin initially occupied by the accreted crustal material. This included, in the area now occupied by the Olympic Mountains, the marine basalt flows that constitute the Crescent formation surrounding the central core of the Olympics. The forearc basin was subject to the ongoing deposition of turbidites and deltaic deposits, which folded and lifted to form the contemporary Coast Range.

The emplacement of the core of the Olympics was accomplished by underthrusting from the west (Figure 4.6b and c). Highly deformed younger marine sediment and volcanic rocks were underthrust eastward beneath the basaltic rocks of the older Crescent formation. At the same time, the Crescent formation was overlaid by sediment along the principal axis of the regional forearc basin. The emplacement of the Olympic core marked the final straightening of the prominent earlier

Figure 4.6 The evolution of the Northwest coast.

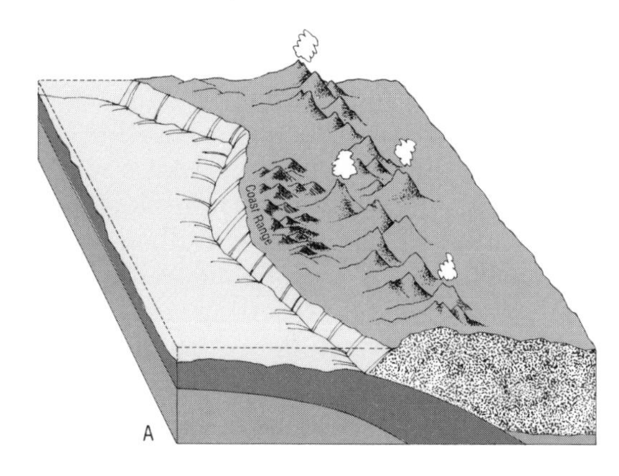

Accumulations of island arc terrane, seamounts, and sediment of continental origin move toward the northwestern edge of the continent and are scraped off the top part of the slab at the subduction zone (120-65 million years before present). As this process continues, there is additional compressional stress and the continent is deformed and uplifted as the coastal range (65-40 million years before present).

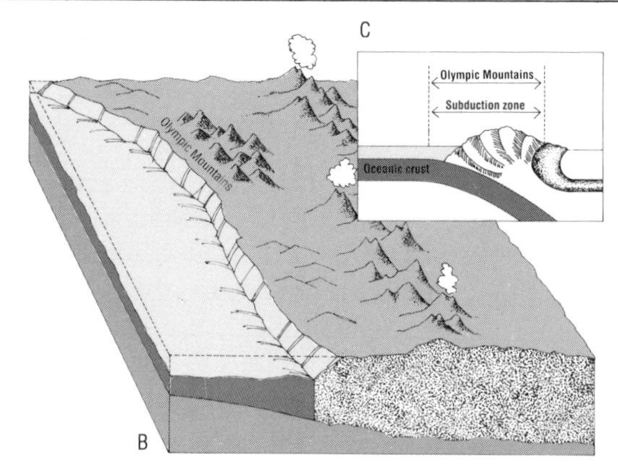

The Western margin is finally "straightened" by the emplacement of the Olympic Mountains. Inset: The Olympics have been formed by a complex convergent boundary process of ongoing underthrusting of material that was originally located on the upper surface of the subducting oceanic plate, but is slowly rotated up and eastward toward the continental mass (40–15 million years before present).

D

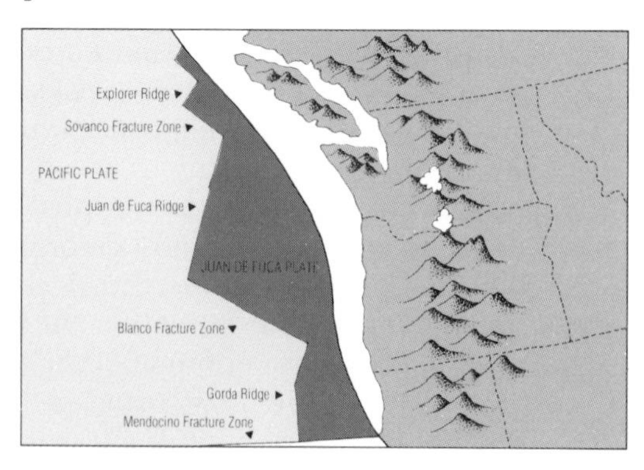

The Northwest Coast as it looks today. The Juan de Fuca Plate (sometimes called the Gorda Plate), being formed at spreading centers along Explorer, Juan de Fuca, and Gorda ridges, continues to be subducted along a convergent boundary with the North American Plate. Shear boundaries are present along the fracture zones.

embayment, and the principal westward extension of the North American continental plate reached the point of today's shoreline.

The Shaping of the Surface by Glaciers

Through all of the many millions of years during which the Cordillera was being developed by global-scale processes, small-scale surficial processes were working and shaping the outermost surface of the crustal rocks. Although the contemporary crustal structure of western North America was established by a few million years ago, the final major shaping of the surface was a relatively recent event. Features resulting from these surficial processes were not mentioned earlier because they are so much smaller in size that their detail is generally obscured by the larger-scale features.

The discussion of this final shaping will require a shift in scales of both size and time: from millions of years to thousands of years, and from continental size to the regional depression known as the Puget Lowland.

Glacial Processes

Glaciers are masses of ice that originate in areas of permanent snowfields and flow down and away from the snowfield. The permanent snowfield may exist because of geographic location (near the earth's poles) or elevation above sea level (in high mountainous regions). In either case the controlling factors are relatively high rate of precipitation and low temperatures. The permanent accumulation of snow results in conversion to ice and eventually high internal pressure results in deformation that causes the ice to flow away from the region of accumulation. Ice flow can range from as high as tens of meters per day to almost imperceptible amounts, but an average rate for contemporary glaciers is not more than a few centimeters per day.

The flow rate of ice is regulated by two factors: thickness of the ice field at the source of the flow, and the slope of the surface over which the ice is moving. Generally, ice flows more rapidly as it gets thicker and as the slope angle increases. If the ice is not flowing, it is *not* a glacier, but simply a subglacial accumulation. Once ice begins to flow, it generally moves out of the area in which conditions favor permanent snowfield accumulation; for example, it may flow downhill into a region of warmer temperatures.

The size of a glacial ice mass depends upon the balance between the rates at which the ice flows and melts. The ice within the body of a glacier is always moving away from the source regardless of what is happening at the terminus. When more ice reaches the foot or terminus of the glacier than is melted away, the terminus advances; when ice melts faster than new ice is supplied, the terminus retreats.

Glaciation is one of the strongest of the surficial geologic processes that shape the surface of the earth. The effects of moving ice are related primarily to the absolute size of glaciers, which (in order of increasing magnitude) are classified in three sizes: valley (or alpine) glaciers, piedmont glaciers, and ice sheets (ice caps).

Valley glaciers originate in a mountain snowfield and follow valleys downslope. Active valley glaciers are still present in the Cascade Mountains and in the Olympic Mountains: Mount Rainier alone has 26 named glaciers.

Piedmont glaciers are formed when several valley glaciers coalesce into a continuous thick sheet of ice. The ice flow originates in the separate snowfields at the heads of the individual valley glaciers, but the flow of the piedmont glacier reflects the larger mass of ice involved. There are no examples of active piedmont glaciers in western Washington but Malaspina Glacier in Alaska is a contemporary example of this class of glaciation.

When a permanent snowfield reaches such a size that it is no longer confined to local depressions, a regional blanket of glacial ice can form. As with any glacier, this ice flows away from the center of maximum accumulation. In contrast to valley or piedmont glaciers, which are confined by topographic features and flow downhill, ice sheet glaciation is capable of overriding local topographic highs. The only ice sheets present today are in Greenland and Antarctica, but during the ice ages climatic conditions gave rise to sheet glaciation much larger than that which exists today.

The process of glaciation yields characteristic features that result from erosion, transport, and deposition of materials. The combination of these features generally provides a strong contrast with adjacent nonglaciated topography.

Although the amount of erosion varies with duration of glaciation and speed of flow, the mere presence of ice accumulation large enough to start glacial flow can cause marked modification in topography. Loose soil and rock are plucked from the surface and frozen in the ice. As the glacier moves, these entrained fragments scour and grind material that they override. In addition to erosion, glaciation results in characteristic landforms of which the most important in the development of Puget Sound is the glaciated valley. In cross-section, glaciated valleys appear U-shaped, rather than V-shaped like stream-cut valleys.

Material deposited by glaciers is called glacial drift—a holdover term from the time when these deposits were attributed to the Biblical flood. Deposited drift can be eroded and redeposited by running meltwater in varying degrees. Drift that has not been reworked by meltwater remains irregularly deposited, poorly sorted material termed *till*. In contrast, layered and better sorted material can result from reworking of the drift by running water (glaciofluvial deposits) or in proglacial lakes (glaciolacustrine deposits).

One of the very special characteristics of the load of soil, rock, and other debris transported by ice is that once frozen into the ice it will remain there until the ice melts. Because most melting is associated with the terminus of the glacier, that is where most material is deposited. As the glacier advances, it may reincorporate earlier deposits and carry them farther. With alternating advances and retreats, earlier deposits may be removed or obscured by later ones.

Another effect of glaciation occurs when the total mass of ice accumulates enough weight to depress the level of the underlying crustal rocks. Subsidence of crustal rocks of the Puget Lowland during extensive regional glaciation resulted in a general local depression of the earth's surface relative to sea level. This depression subsequently rebounded after the ice retreated from the region, which was a factor in the shaping of Puget Sound.

Cordilleran Glaciation in Western North America

Although geological evidence indicates that there have been several periods of major global glaciation, it was the most recent—during the last several tens of thousands of years—that was the principal agent in establishing Puget Sound as the dominant feature of the Puget Lowland. Prior to the glacial period the Puget Lowland was part of a broad regional depression extending from the lower Fraser River Valley southward through the Willamette Lowland to the northern slopes of the Klamath Mountains. Instead of being flooded, the area now occupied by Puget Sound was composed of piedmont and alluvial plains drained by relatively minor rivers originating principally on the west slope of the Cascades. These rivers drained northward and joined the Fraser discharge emptying into an ancestral Strait of Juan de Fuca. The situation was somewhat analogous to today's drainage of the Willamette Lowland into the Columbia River.

Global glaciation during the past few tens of thousands of years was characterized by four major advances of glacial ice moving across northern North America and Europe. At its maximum extent, the ice covered about one-third of the continent and the terminus advanced and retreated as climatic conditions changed. The major source of glacial ice on North America was the Laurentide Ice Sheet (Figure 4.7), which was centered over Hudson Bay; but the western portion of the continent was affected by the Cordilleran Ice Sheet.

As climatic cooling occurred, marking the beginning of the glacial age, the permanent snowline crept down to lower elevations in the mountains along the Cordillera. The result was a marked increase in valley glaciation, extending far south along the crest of the Cascades.

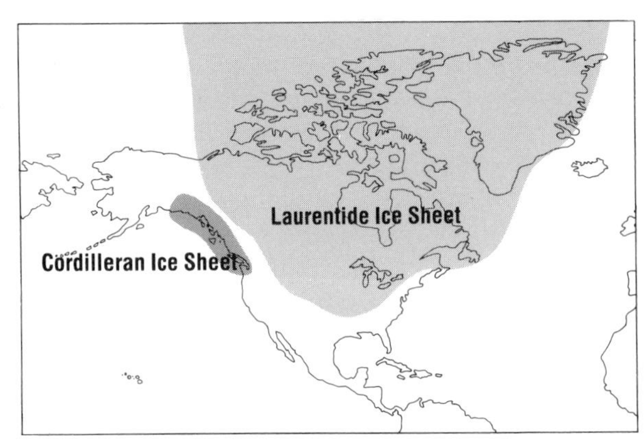

Figure 4.7 The extent of the Laurentide and Cordilleran ice sheets.

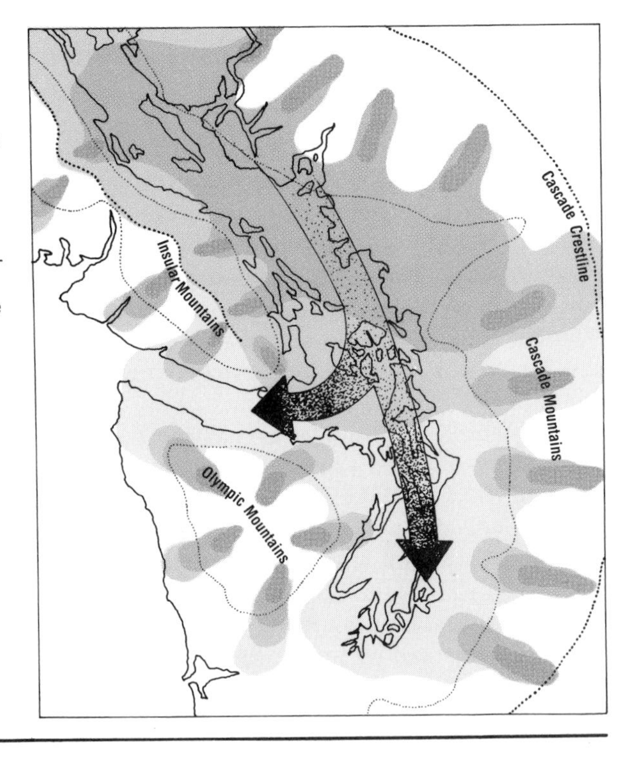

Figure 4.8 Growth of the Puget Lobe. With climatic cooling, valley glaciers grew and moved down the slopes of the Cascade and Olympic mountains, and the Insular and Coast mountains of British Columbia. As the smaller glaciers coalesced, they formed a larger mass of ice that moved south into the Puget Lowland and west out the Strait of Juan de Fuca.

To the north, on the western slopes of the Coast Mountains of British Columbia, rapidly accumulating ice along the valleys coalesced and developed a prominent piedmont glacier (Figure 4.8). It generally flowed to the west and the terminus eventually calved into the Pacific Ocean. However, at the southern end, the normal downhill westward flow was blocked by the Insular Mountains on Vancouver Island. Joined by additional ice from the east slope of the Insular Mountains, a mass of ice grew, centered over a topographic depression in the Strait of Georgia. This ice eventually began to flow laterally between Vancouver Island and the mainland. A large portion flowed southward with a mass so great that it overrode the San Juan and Gulf Islands.

Upon reaching the Olympics, which were high enough to block and divert the flow, the ice split into two separate lobes. One portion, the Juan de Fuca Lobe, moved westward along the depression between the north flank of the Olympics and the southern flank of the Insular Mountains. The other portion, the Puget Lobe, moved southward into the Puget Lowland. These two lobes were near the terminus of the glacier and advanced and retreated several times. At least four major advances of the Puget Lobe have been identified that correspond roughly with the four stages of glaciation constituting the late Cenozoic ice ages.

The erosive power of the local Cordilleran glaciation was strongest at the north; glacial deepening occurred in Haro Strait, East Sound, and Rosario Strait in the San Juan Islands. The glacier also carved grooves and scoured the highest peaks in the islands. Within the Puget Lowland, the main channels of the Puget Sound system were downcut by glacial erosion and modified by glacial deposition. Most glaciated portions of the Puget Lowland are covered by glacial deposits that date from the retreat of the last glaciation. Records from earlier ice advances and interglacial stages are extremely sparse in the Puget Lowland and are primarily found southeast of Puget Sound, in an area that was not severely affected by the final advance of the Puget Lobe.

Local Glacial Chronology

The several advances and retreats of the Puget Lobe have been reconstructed by glacial geologists on the basis of type, extent, and superposition of the glacial features and deposits of the Puget Lowland and the flanking slopes of the Olympics and Cascades (Table 4.1). Characteristically, deposits associated with the Puget Lobe contain rock and debris of *northern provenance*, igneous and metamorphic rocks originating in the Coast Mountains of British Columbia and the North Cascades. The presence of these rocks in the Puget Lowland implies glacial transport into the region from the north and is attributed to Puget Lobe activity. In contrast, a different provenance can be used to identify deposits originating from valley glaciation on the western flanks of the Middle Cascades and, to a more limited extent, the eastern slopes of the Olympics.

The earliest identifiable advance of the Puget Lobe is defined by northern provenance glacial till found in the Green River Valley just west of the Cascade front, and along the east side of the Puyallup River near Orting, Washington (Figure 4.9). The northern provenance till at

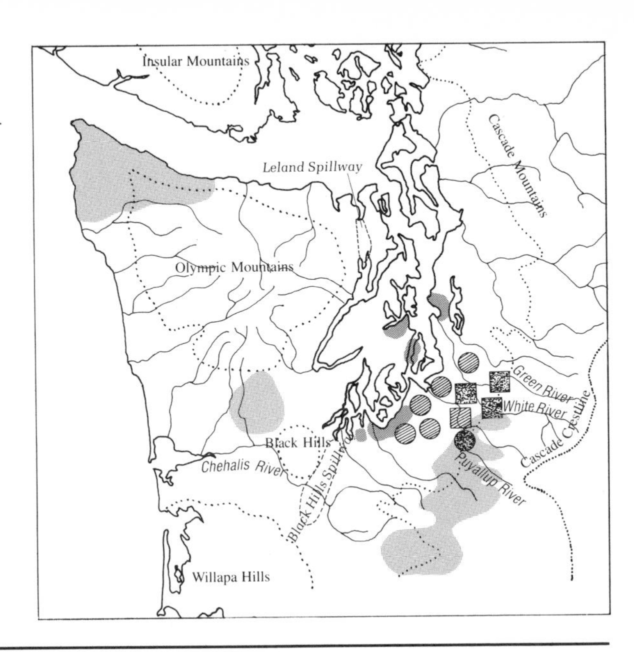

Figure 4.9a Areas with exposures of pre-Fraser glacial and interglacial deposits.

Table 4.1 Sequence of glacial and interglacial events. Ages of earliest events are less certain than those of more recent ones.

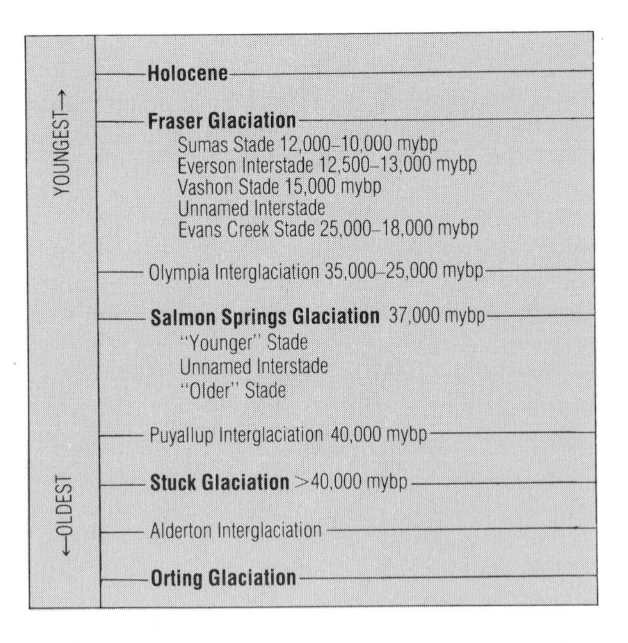

	Event
YOUNGEST→	**Holocene**
	Fraser Glaciation
	Sumas Stade 12,000–10,000 mybp
	Everson Interstade 12,500–13,000 mybp
	Vashon Stade 15,000 mybp
	Unnamed Interstade
	Evans Creek Stade 25,000–18,000 mybp
	Olympia Interglaciation 35,000–25,000 mybp
	Salmon Springs Glaciation 37,000 mybp
	"Younger" Stade
	Unnamed Interstade
	"Older" Stade
	Puyallup Interglaciation 40,000 mybp
	Stuck Glaciation >40,000 mybp
	Alderton Interglaciation
←OLDEST	**Orting Glaciation**

Orting is interbedded with till of Cascade provenance implying oscillation of the terminus of the Puget Lobe. This initial glaciation has been named after the type section at Orting.

Following the Orting glaciation the Puget Lobe retreated back to the north, during a period known as the Alderton interglaciation. Dur-

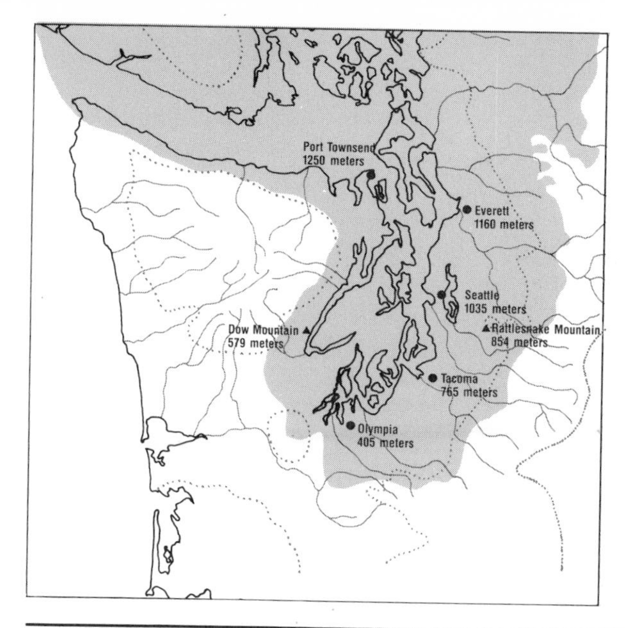

Figure 4.9b Maximum extent of Cordilleran Ice during the Vashon Stade of the Fraser glaciation.

Extent of mountains

▲ Nunataks, showing thickness of ice

● Representative thicknesses of ice

ing this time nonglacial materials were deposited on top of glacial deposits of the Orting age, and can be seen exposed along the Puyallup River near the town of Alderton. The Alderton nonglacial material provides a general indication of a return from glacial to interglacial conditions. The presence of pollen, for example, indicates flora existed that had temperature requirements comparable to today's climate. Further, there is mineralogical evidence that implies a period of active volcanism in the adjoining Cascades.

The Stuck glaciation takes its name from a now archaic appellation for the lower White River just north of the town of Sumner. Glacial till in this area and in the valley walls of the Puyallup and the Green rivers has northern provenance and implies the return of the Puget Lobe. The till is sometimes interbedded between glaciofluvial deposits, sometimes has glaciolacustrine interbeds; both conditions imply fluctuations of the terminus of the ice. These relatively minor fluctuations within a period of glaciation are termed *stades* (when the terminus advances) and *interstades* (when the terminus retreats).

Accumulations of stream-deposited sands and gravel, lacustrine clays, and peat bogs dating from a period of interglaciation are found in the southeastern Puget Lowland. The peat bogs and pollen collected from these deposits imply climatic conditions similar to today. Mount Rainier was active during this period and both volcanic ash and mudflows are part of the interglacial deposits. Locally, this period is termed the Puyallup interglaciation. It may correlate relatively closely with the Sangamon interglaciation of central North America and can be dated as older than about 40 thousand years before present.

Two advances of the Puget Lobe have been identified as stades in the Salmon Springs glaciation. These are separated by an interstade during which the terminus of the Puget Lobe retreated north of the present position of Seattle. During the interstade, marine conditions prevailed in flooded portions of the glacially scoured lowland indicating retreat of the Puget Lobe far enough to permit open connection with the Strait of Juan de Fuca and the Pacific Ocean. Peat samples from this interstade have been dated as older than 38 thousand years before present and the climate during the short period of glacial recession was cooler and moister than it is now. The maximum extent of the Puget Lobe during Salmon Springs glaciation is obscured by later glacial activity, except in the southwest Puget Lowland where some weathered glacial drift of Salmon Springs age is found beyond the limit of later drift in the area east of the Satsop River about 25 kilometers west of Shelton.

The Olympia interglaciation was a return to interglacial conditions within the Puget Lowland. The period extended from about 35 to 15 thousand years before present, during which the lowland was characterized by shallow lakes, swamps, and some forest cover of pine and spruce. Drainage was northward into the Strait of Juan de Fuca. The climate, considering the vegetation, was at least as moist as present and probably somewhat cooler. Deposits from this interglacial period are stream and lake sediments, which are found along the shore of southern Puget Sound between Tacoma and Olympia, the south shore of Sinclair Inlet, and Colvos Passage. A prominent section of these interglacial deposits is exposed in a section about 20-meters (70-feet) thick in a sea cliff at Magnolia Bluff in Seattle.

Onset of the Fraser glaciation was marked by a brief period of increased valley glaciation. This, the Evans Creek Stade, affected the uplands on the western slope of the Cascades but had only an indirect effect on the lowlands, principally by modifying the type of sediment being carried to them. A brief interstade was followed by the Vashon Stade, the most recent of the southerly incursions of the Puget Lobe. This may or may not have been the most massive of the local glaciations. However, in keeping with the principle that the "most recent leaves the most evidence," the Vashon glacial deposits are the most widespread and the details of the Vashon are best understood. The Puget Lobe of the Vashon Stade began its advance onto the Puget Lowland and by about 18 thousand years before present had formed an ice-dam at the north, blocking the normal northward drainage. As the ice continued to advance southward a glacial front lake was formed by impounded runoff and meltwater. Sediments deposited in the lake (glaciolacustrine) covered much of the area, overlying the nonglacial sediments that had accumulated previously on the lowland. There is, however, very limited direct evidence for these lake sediments since most of them were subsequently overridden and reworked by the advancing ice. With drainage to the north blocked by the ice-dam, the lake water overflowed to the south.

The principal outlet for the overflow was along a route extending from the southern end of what is now Budd Inlet southwestward between the Black Hills and Tumwater, via Black Lake and the Black River valley, and into the Pacific westward through the Chehalis Valley (Figure 4.9). This route is called the "Black Hills Spillway," and the water from this and other outlets converged on the Chehalis Valley. With a very considerable volume of water to be drained, the Chehalis Valley was developed as a much larger valley than would have been formed by the small river that occupies it today.

The details of the general direction of ice flow in the Puget Lobe during the Vashon glaciation are reflected in several scales of topographic features in the Puget Lowland. The prominent channels occupied by Possession Sound, the Main Basin of Puget Sound, and Dabob Bay-Hood Canal, were scoured deeply into the pre-existing sedimentary deposits of the lowlands. In addition, many smaller topographic features display lineations in the glacial deposits, which indicate the direction of the ice flow. These lineations are common over much of today's lowland and are particularly well developed in the region between Mason Lake and Pickering Passage, on Reach and Stretch Islands, on parts of Hartstene Island, and over many parts of the Kitsap Peninsula.

When the ice reached its maximum extent, about 14 thousand years ago, the Puget Lobe was laterally constrained between the mountain fronts of the Olympic and Cascade Mountains. During this period of maximum southward extent, the thickness of the ice was measured in hundreds of meters, estimated by determining the height on the flanking mountains below which there is evidence of glaciation and above which there is none. For the Puget Lobe, one estimate of this height can be made by determining the maximum elevation at which glacial *erratics* are found. These erratics are rocks which are lithologically distinctly different from the local bedrock, and which could only have been carried to their present site by glaciers. In this instance, lighter colored granitic rocks from the Coast Mountains of British Columbia contrast with the darker colored rocks from the Northern Cascades or the Olympics.

Another estimate can be made by determining the elevation above which glacial erosion or scour is absent. This latter technique is particularly useful in the case of *nunataks*, the isolated high peaks which project above the level of a surrounding ice field. At the time of maximum advance, only two nunataks are known to have projected upward through the ice. These are Dow Mountain just east of Lake Cushman and Rattlesnake Mountain southwest of North Bend. Both of these were relatively close to the lateral margins of the ice and, coincidentally, at almost the exact same latitude.

Based on data of this kind, the thickness of the main glacial lobe is estimated to have been about 1,600 meters (5,250 feet) over the present site of Bellingham, and about 1,200 and 1,300 meters (3,950 and 4,250 feet) where it separated into the Juan de Fuca Lobe and Puget Lobe. The Puget Lobe was about 1,000 meters (3,275 feet) thick over Seattle's location, decreasing to 700 to 800 meters (2,300 to 2,625 feet) over Tacoma and about 450 meters (1,475 feet) over Olympia.

The maximum southerly extent of the Vashon ice was a terminus that extended across the lowland south of Olympia. The position of this maximum advance has been determined by a combination of morainal and recessional deposits which gives rise to a characteristic local topography, an irregular and nonsystematic arrangement of many hummocks and depressions. Hills called *kames* are small conical or irregular ridges composed of gravel and sand formed by sediment from streams flowing off the glacial terminus and deposited against the ice front. They are one of several ice-contact features and display some degree of layering or stratification. They are frequently associated with terraces (kame terraces) built up by glacier-derived sediments deposited between the ice front and either the walls of adjoining valleys or higher portions of terminal moraine. *Kettles*, irregular depressions in the land surface, resulted from wasting of ice blocks that had been buried, either wholly or partially, in the ice front sediment deposits. In combination these features form the distinctive local topography associated with the maximum southerly advance and initial recession of the Vashon ice. One of the locales where these features may be observed is in the area south and east of Olympia. Ward Lake, Hewitt Lake, and Smith Lake are characteristic kettles, and the region around Long Lake displays a *kettle and kame* topography.

Along its lateral margins, the ice of the Puget Lobe eroded prominent truncations across the east-west valleys and ridges of the alpine-glaciated Olympic and Cascade Mountains. As the Puget Lobe began to recede, sands and gravels from the flanks of the regional glaciers and from the smaller glaciers in the mountains formed prominent deposits of glacial outwash. Several large depositional plains remain south of Tacoma and Olympia, and west of Shelton. Outwash drainage channels extend westward and southwestward out of the Cascades and southward from the Olympics to the Chehalis Valley.

For some time after the ice began to recede, the ice-dam to the north precluded drainage northward and a complex sequence of glacial lakes occupied the uncovering Puget Lowland. As the Vashon ice retreated, the lakes evolved in size and location with successively lower outlets and changing interconnections. The more prominent of these occupied the larger channels and valleys that were scoured by the Puget Lobe and were the principal sites for glaciolacustrine deposition of layered silts and clays as well as delta deposits associated with the principal streams discharging into the lakes. During much of the recessional period lakes continued to drain through the Black Hills Spillway to the south. To the north, although the main ice-dam of the Puget Lobe persisted, the Juan de Fuca Lobe retreated and open marine conditions were re-established in the strait.

As the Puget Lobe continued to retreat, initial clearing of the ice-dam was first accomplished at the northwest. There was a short period when, although the southerly drainage for the glacial lakes continued, a northerly drainage into the Strait of Juan de Fuca was established for glacial Lake Leland, which occupied the Hood Canal-Dabob Bay area. This outlet was through a channel between Dabob Bay and Port Discovery via contemporary Lake Leland. It was at a lower elevation than the overflow channels of the lakes to the south. The southward drainage was successively abandoned in favor of northward drainage, initially via the Leland Spillway and ultimately through Admiralty Inlet, when the Puget Lobe retreated far enough north to complete the removal of the ice-dam.

The Vashon ice of the Puget Lobe retreated northward from the Puget Lowland about 13 thousand years ago. Although a later advance of the Fraser glaciation during the Sumas Stade affected the Fraser River Valley, the Vashon glaciation was the last incursion of ice onto the Puget Lowland. Some limited glaciation has continued until today in the Olympic and Cascade mountains but almost all of the prominent features of the surface topography of the lowlands were established by the close of the Vashon Stade.

CHAPTER FIVE

The Shape of Contemporary Puget Sound Environs

To a geologist attempting to develop an understanding of the region's physical origins the discussion so far has perhaps put the cart before the horse. Historically, description and characterization of landforms normally precedes speculation about how they were shaped. Instead, having already outlined the geological processes and history of the evolution of the local region, we will now look in more detail at the form of local topography.

Individual perspective frequently results in very different perceptions of what is viewed. An astronaut, viewing Puget Sound from a high altitude, sees a long, narrow, inland sea. A sailor, on the other hand, might view Puget Sound as a relatively complex maze of interconnected waterways. To find his way around within the system he might use bearings on prominent features—such as high peaks in the adjoining mountains, or distinctive landmarks along the shore of the lowlands and islands. To avoid running his ship aground or on a rocky shoal he would also be concerned with the shape of the seafloor, which he normally could not see. All of these features are topographic forms comprising a continuum that shapes the earth's surface on land and under the sea.

The Flanking Mountains

The Olympic Mountains

The Olympic Mountains (Figure 5.1), which dominate the central portion of the Olympic Peninsula to the west of Puget Sound, are a part of the coastal mountains of the western United States. The lowest elevations of the coastal mountains are in southern Washington and Oregon (a tendency also observable in the Cascades) and the Olympic Mountains rise to the highest peak elevations along the whole coastal belt. These higher elevations appear even more pronounced because of their isolation from other parts of the mountains that border the Pacific Coast, but they are not as high as Cascade peaks. They are separated from the Oregon Coast Range by the Chehalis River Valley and Willapa Hills, and isolated from the Insular Mountains of Vancouver Island by the Strait of Juan de Fuca.

The highest elevation in the Olympic Mountains is 2,424 meters

(7,954 feet) at Mount Olympus, but there are several other peaks with elevations exceeding 2,300 meters (7,000 feet). Occupying an area of about 10,000 square kilometers (4,000 square miles), the higher areas are extremely rugged and dissected by a complex system of valleys. Local relief frequently exceeds 1,000 meters (3,000 feet) between high crests and adjacent valley floors, and slopes of less than 20 percent are uncommon. The general crest elevations are highest near the center of the Olympic Mountains, with ridges and peaks in excess of 1,800 meters (6,000 feet). This general crest height decreases radially away from the center to elevations of about 600 meters (2,000 feet) and then descends steeply. The surrounding lowland areas are composed of a narrow band approximately 15 kilometers wide (10 miles) of low hills on the north and west, and a broader band on the south that grades into the valley of the Chehalis River. On the east, the descent into the Puget Lowland and sea level at Hood Canal is very abrupt.

Stream and glacial erosion of the uplands has resulted in a disordered arrangement of peaks, jagged ridges, and steep-sided valleys. There is a pattern of radial drainage imposed by prominent streams that originate in glaciers or snowfields in the higher interior then flow outward from the center. Much of the rough topography resulted from erosion by streams and there is an increasing level of dissection which generally increases outward from the center of the Olympics. Because of the extremely rugged topography combined with dense vegetation, particularly on the western slopes, the region has always been very sparsely populated.

Much of the smaller scale topography in the Olympic Mountains is the result of glaciers and glacially related processes. In spite of the low elevation of the Olympics compared to the Cascades, abundant precipitation results in snowfield accumulation that is heavy enough to survive the region's mild summer season. These local conditions favor development of alpine glaciers, and the Olympic Mountains contain about 50 permanent snowfields with glacier characteristics. Generally located on northern slopes at elevations above 1,220 meters (4,000 feet), the most prominent of these snowfields are on and near Mount Olympus. The longest one is the Hoh Glacier at Mount Olympus; it begins at about 1,675 meters (5,500 feet) elevation, extends about 13 kilometers (8 miles) down the flank of the mountain and terminates at the 1,220-meter level (4,000 feet).

A secondary group of snowfields is centered around Mount Anderson. Here, Eel and Anderson glaciers are formed at somewhat higher elevation—above 1,820 meters (6,000 feet)—than at Mount Olympus, flow for a shorter distance, and terminate at about 1,675 meters (5,000 feet) elevation. Not all of these snowfields are true glaciers (i.e., have glacial ice flow rather than simple accumulation).

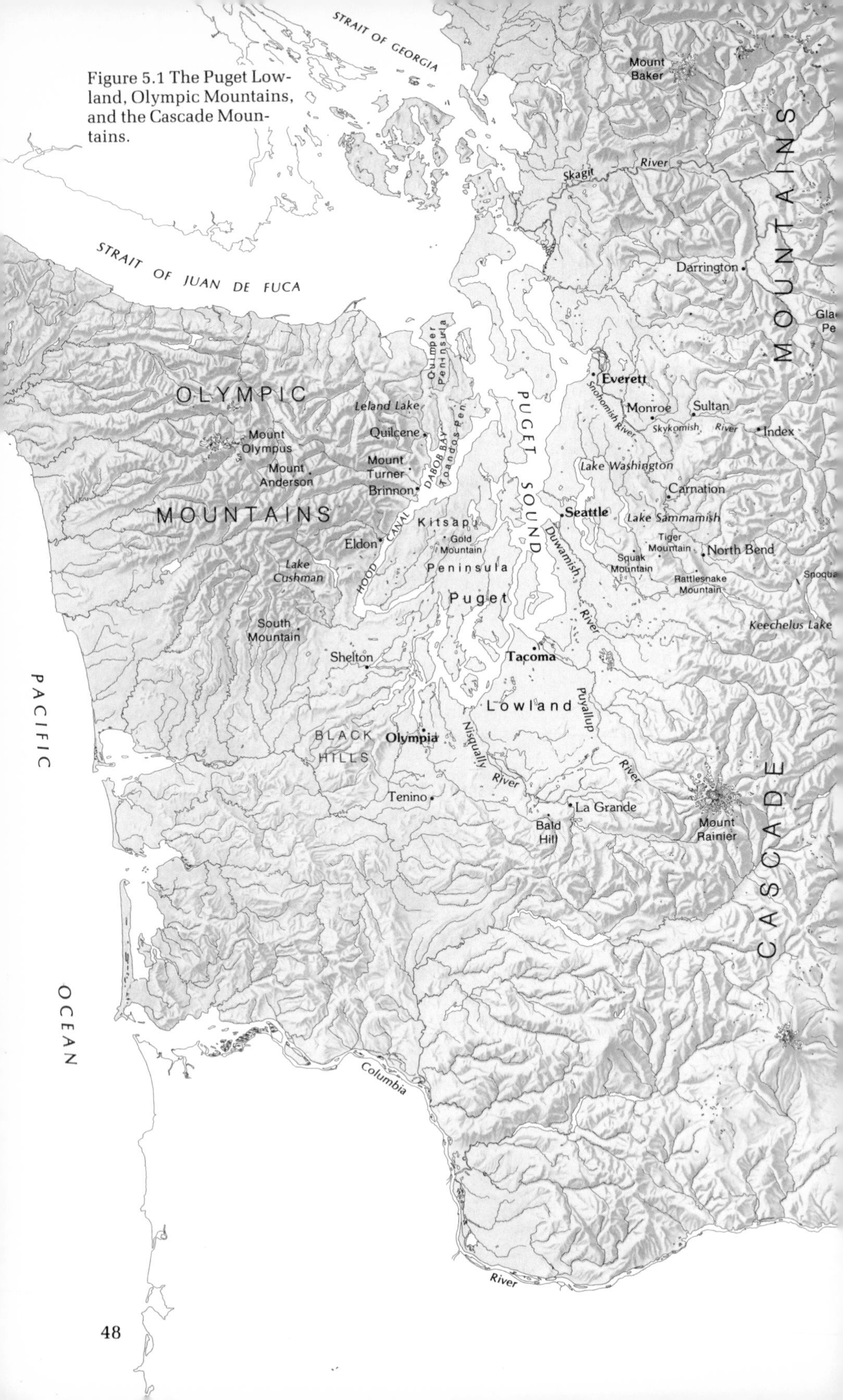

Figure 5.1 The Puget Lowland, Olympic Mountains, and the Cascade Mountains.

Because of extremely steep terrain, visits to many snowfields are limited to helicopter landings or on-foot exploration by some of the more dedicated hikers and climbers who visit the area. From evidence obtained directly from the snowfields that have been explored, and indirectly from aerial photographs, it appears that the Olympic glaciers are receding.

A number of alpine lakes occupy abandoned glacial sources. These lakes are characteristic of regions once subjected to alpine glaciation, and occur when water collects in a *cirque*, the depression eroded at the source of a former glacier. Some examples, such as Claywood Lake and Mildred Lakes, are headwater feeders, respectively, for tributaries of the Dosewallips River and the Hamma Hamma River.

The U-shaped valleys associated with alpine glaciation can be observed at many localities. Some prominent examples adjoining the Puget Lowland are the valleys at the Hamma Hamma and Duckabush rivers. An interesting example of the interplay between Olympic Mountain alpine glaciation and Puget Lobe regional glaciation can be observed at Lake Cushman. Here, ice of the Puget Lobe extended westward around Dow Mountain and moved upward along the course of the ancestral North Fork Skokomish River. The portion of the valley now occupied by the western part of Lake Cushman was overdeepened by glacial scour; the lake itself was formed and is maintained by damming caused by morainal deposits from the main glacier. The present, almost right angle, diversion of the North Fork southward to join the Skokomish was established by the Puget Lobe ice front blocking eastward stream flow and diverting it southward along the ice margin.

The Cascade Mountains

The crestline of the Cascade Mountains is the dominant feature establishing the easternmost limit of the drainage basin of Puget Sound (Figure 5.1). Culturally and historically the Cascades have had profound political and economic impact upon the Puget Sound region because of the constraints they have imposed upon east-west travel and transportation.

By name and morphologic continuity the Cascade Mountains extend southward from the Fraser River Valley of British Columbia, through Washington and Oregon, and connect with the Sierra Nevada near Mount Lassen in California. The portion of the Cascades that forms the eastern slope of the Puget Sound drainage basin may be subdivided at about the latitude of Snoqualmie Pass. This separation into North and Middle Cascade Mountains (the South Cascades, a further subdivision, lies much farther south than the local area) is based upon several contrasting characteristics.

All of the Cascade range is a high mountain regime with local relief (valley floor to adjoining peak crest) characteristically in excess of 1,000 meters (3,000 feet). Much of the terrain is extremely rugged and slopes are very steep almost everywhere. The general crest height is remarkably uniform over much of the Cascades, and is exceeded at specific sites only by the peaks of the volcanoes: Mount Baker (3,285 meters; 10,778 feet), Glacier Peak (3,213 meters; 10,436 feet), Mount Rainier (4,405 meters; 14,410 feet).

On a time scale that is perhaps comprehensible (thousands of years) the volcanic peaks provide a good example of ongoing geological processes that first raise and then lower the surface of the earth. Each of the three local peaks was formed by volcanic activity resulting in a volcanic peak that was built up above the adjacent land surface. Initial development of these peaks took place during major regional glaciation. Alpine glaciation continues at these peaks, which are sites of much of the remaining glacial activity in the Cascades. Consequently, their history has been an alternation of building during periods of active volcanism and downcutting by glacial erosion during dormant periods.

The North Cascades display a more rugged terrain than the Middle Cascades south of Snoqualmie Pass. This contrast is due in large part to the different ways in which glaciation developed today's surface landform. Prior to the last period of major glaciation of 10 to 50 thousand years ago, the Cascade uplands were considerably more regular than now. The regional uniformity in the present crestline of the Cascades is related to the preglacial upland into which the glaciers cut deeply. This glacial downcutting was not directly related to the action of the Puget Lobe but was caused by alpine glaciation on the higher flanks of the upland east of the Puget Lobe. In the North Cascades, alpine glaciation was both more intense and persistent than in the Middle Cascades, and the contrast in today's topography is related to the degree of glacial dissection of the preglacial upland. The northern area was subject to a greater density (number) of alpine glaciers than the south, where glacial action tended to be less dense and effects more localized.

As regional glaciation waned, alpine glaciation in the uplands also diminished. Today, the few remaining glaciers are generally associated with the higher elevations of the volcanic peaks (e.g., Mount Baker, Glacier Peak, Mount Rainier) where they tend to radiate outward and downslope from the highest elevation near the peak. Other active glaciers are associated with north facing slopes at high altitudes near the crestline.

The extensive glaciation of the Cascade Mountains has left its mark on many of the local landforms. In addition to the general dissection of the western slope of the uplands, sites of former glaciers are abundant and can be identified in many places by the presence of small alpine lakes. These are most obvious near the crestline, but some occur at lower elevations. Probably the greatest concentration of alpine lakes is associated with the sources of upland tributaries of the Middle Fork of the Snoqualmie River.

U-shaped valleys formed by alpine glaciation are also abundant. Many of the large streams that drain the west flank of the Cascades occupy former glacial valleys. One easily observable example is the South Fork of the Snoqualmie River along highway I-90 westward from Snoqualmie Pass. Farther north, the valley of the North Fork of the Skykomish River northeast of the town of Index is another prominent example.

The best examples of moraine-dammed lakes are just to the east of the Cascade crestline (e.g., Keechelus Lake). On the west flank, several of the lakes that feed the North Fork Snoqualmie River north of North Bend are further examples of such lakes.

At lower elevations many rivers appear to be out of place in the valleys they occupy. The stream course meanders on a valley floor that is even more flat-bottomed than the U-shaped valleys at higher elevations. This additional flattening is frequently the result of deposition of glacier derived sediment during later stages of receding glaciation and the present river is flowing over these deposits. Some examples of this may be observed on the Sauk River northeast of Darrington, on the Skykomish River east of Sultan, and on the Snoqualmie River south of Carnation.

The most prominent landform of the Cascade Mountains is the volcanic peak. In sharp contrast with the rest of the Cascades, these are isolated peaks rising well above the crestline. In the sector of the Cascades that forms the eastern watershed for the Puget Sound drainage basin, Mount Baker, Glacier Peak, and Mount Rainier are three of a larger group of similar peaks that form a characteristic feature of the Cascades along their long north-south extent. These volcanoes were formed very late in the development of the region and were a result of Andean-type convergent boundary processes, which are still active locally.

The Puget Lowland

The Puget Lowland (Figure 5.1) is the local subdivision of a regional morphologic depression extending from British Columbia southward into Oregon. This regional feature extends northward from the Puget Lowland into the Fraser Lowland, which includes the southern portion of the Strait of Georgia. To the south it includes the valleys of the Chehalis, Cowlitz, and lower Columbia rivers, extending still farther south to include the Willamette Lowland in Oregon. In the larger sense, the regional lowland lies between the Coast Ranges to the west and the Cascade Mountains to the east. Locally, the Puget Lowland lies between the Olympic Mountains and the North and Middle Cascade Mountains.

The Puget Lowland has been subjected to regional glaciation and the contemporary landscape is predominantly the result of erosion and deposition associated with the incursion of the Puget Lobe during the ice-ages. It contains very few exposures of older rock—unlike the Olympic and Cascade Mountains—and is almost entirely covered by glacial debris. The result of this glacial deposition is a subdued relief contrasting sharply with that of the rugged flanking mountains.

In a general sense, the land area of the Puget Lowland slopes downward from the flanking mountains and hills along its southern border. Most of the lower elevations are covered with glacial till (a very poorly sorted accumulation of clay- to boulder-sized fragments), and the topography consists of gently rolling or hilly ground.

The Southern Puget Lowland

The southern limit of the Puget Lowland is the drainage separation between the streams entering Puget Sound and those flowing into the Chehalis River. This boundary approximates the southernmost extent of the Puget Lobe glaciation, which either terminated against preexisting topographic highs (such as the Black Hills) or left behind a series of morainal hills or outwash. The exact limits of this southern boundary are not always too obvious. There is a discontinuous belt extending from southwest of Shelton to the lower slopes of the Cascades near La Grande. At the Black Hills and to the east of Tenino, where there are elevations in excess of 760 meters (2,500 feet), the hills are obvious and it is reasonably clear where the ice front terminated. Local relief ranges from 45 to 425 meters (150 to 1,400 feet) along the east-west trend of this region.

West and immediately east of the Black Hills the Puget Lowland boundary is less well defined. These were sites of the principal overflow channels for the lake water impounded south of the Puget Lobe. Local relief is much more subdued, averaging only 30 to 60 meters (100 to 200 feet), and the topography is characterized by the very flat"prairie"topography such as that located between South Mountain and the Black Hills, and the former Black Lake Spillway between the Black Hills and Tenino.

Southeast of Tacoma are extensive outwash plains similar to the prairies associated with the Black Lake Spillway. These plains are underlaid with sandy gravel, and small lakes and marshes abound.

The Western Puget Lowland

The western boundary of the Puget Lowland is defined by a relatively abrupt change in slope along the eastern limits of the Olympic

Mountains. Some of the steepest gradients in the whole region are along this boundary. For example, the land surface slopes downward from elevations greater than 910 meters (3,000 feet) at Mount Turner to depths of more than 180 meters (600 feet) in Dabob Bay in a horizontal distance of about 10 kilometers (4 miles). Comparable changes in elevation over short distances characterize the western boundary between Eldon and Brinnon, and there is nothing comparable to these gradients anywhere else along the margins of the Puget Lowland.

The western portion of the Puget Lowland may be divided into three sectors: the land west of Hood Canal and Admiralty Inlet, Hood Canal itself, and the Kitsap Peninsula. The land along the western boundary includes a narrow strip of lowland west of Hood Canal and the area east of Leland Lake.

The narrow strip of Lowland at the west extends northward paralleling Hood Canal. Widest (about 15 kilometers, 10 miles) at the southwest between South Mountain and Hood Canal, it narrows northward and is pinched out between Brinnon and Quilcene, where Mount Turner and Mount Walker are almost adjacent to the shore of Dabob Bay. The relief in this sector of the lowland is of the order of a few tens of meters (a few hundred feet) with a gentle slope, increasing in elevation from sea level at Hood Canal, westward to the sharp change in gradient marking the boundary between the lowlands and the Olympic Mountains. At the south, the North Fork Skokomish River flows south from Lake Cushman, in a valley with gently sloping flanks, across the broadest portion of this part of the lowlands. Northward, as the strip narrows, the limit of the lowlands extends westward along the lower valleys of the eastward flowing Hamma Hamma, Duckabush, and Dosewallips rivers.

Geographically, the land included in the Toandos Peninsula northward to the Quimper Peninsula is part of the western portion of the Puget Lowland. The western boundary of the lowland is still defined by the break in slope upward from the lowlands into the eastern uplands of the Olympic Mountains. This boundary is locally west of Leland Lake and this portion of the lowland is much broader in east-west extent than the narrow strip along Hood Canal. Composed principally of glacial till, this is an area of moderate relief of a few hundred feet with maximum elevations approaching 185 meters (600 feet) above sea level.

The Kitsap Peninsula occupies the western portion of the Puget Lowland between Hood Canal and the Main Basin of Puget Sound. Once a group of islands in the postglacial lake, it is now connected to the mainland across the former Clifton Channel between the head of Hood Canal and the head of Case Inlet. The peninsula is less than 150 meters (500 feet) elevation, except in uplands west of Bremerton and in a series of north-south elongated hills along the southwestern shoreline. The prominent uplands west of Bremerton contain one of the rare outcrops of nonglacial bedrock in this portion of the lowland. At the adjacent Green Mountain and Gold Mountain, peak elevations are in excess of 485 meters (1,600 feet) above sea level, and preglacial bedrock is exposed. Although there is no evidence that these mountains were nunataks (projected above the level of the Puget Lobe), they apparently played a strong role in modifying the general flow of the Puget Lobe along its western margin. The landscape over most of the Kitsap Peninsula consists of low relief hills, composed of glacial till, which display characteristic elongations parallel to the direction of ice flow of the Puget Lobe. The axis of elongation is predominantly north-northeast to south-southwest, a trend that is also reflected in many streams (e.g., Tahuya River) and by the linear arrangement of many small lakes that occupy a former outwash channel to the west of the Tahuya River.

The Eastern Puget Lowland

The eastern boundary of the Puget Lowland is not as easily defined as either the southern or western boundaries. The eastern limit of the Puget Sound drainage basin is the crestline of the Cascades. The western flank of the Cascades slopes downward toward the lowland but much of the lower elevations were subjected to the effects of the Puget Lobe as well as more localized alpine glaciation. This resulted in an extremely irregular boundary, characterized by prominent eastward incursions of lowland along the major river valleys, separated by westward extensions of upland that are morphologically considered part of the Cascades. Westernmost encroachment of the Cascades onto the lowland contains exposures of preglacial bedrock and is marked by a series of peaks, some higher than 610 meters (2,000 feet). These peaks extend westward from Rattlesnake Mountain, including Tiger and Squak mountains south of Lake Sammamish.

Much of the lowland eastward from Puget Sound displays the characteristic topography of glacial till deposits. The elevation is, with some local exceptions, less than 150 meters (500 feet) above sea level. Much of the landscape consists of rolling hills with a relief of only a few hundred feet. In contrast with the Kitsap Peninsula, lineations in relief features are not as common, but some obvious examples exist. Lake Washington and Lake Sammamish occupy two prominent linear depressions in this part of the lowland.

Drainage patterns of preglacial streams was severely modified by the regional glaciation. During the period of active glaciation and ice-damming, many streams originating in the Cascades cut new channels. When the glaciation receded, some of these channels were maintained, some were abandoned as streams reverted to the former course, and some new channels were established. Although active downcutting of valleys continues in the uplands, the streams flowing across the eastern Puget Lowland are running across low gradient, wide, flat-floored valleys filled with glacial and alluvial debris. This condition is relatively unstable since the streams are not confined to clearly defined valleys.

Many rivers enter Puget Sound via a complex system of distributaries. At Everett, for example, the Snohomish River enters Possession Sound through distributaries (sloughs) that have shifted position and relative importance in recent historical time. Comparable conditions prevail at the mouths of the Skagit and Stillaguamish rivers and, to a lesser degree, the other major rivers entering Puget Sound from the east.

Farther upstream these rivers still run through broad valleys and frequently pursue an apparently aimless course as they leisurely move seaward. Examples of this may be observed on the Skagit River east of Sedro-Woolley, and on the Snoqualmie River between Duvall and Carnation.

Flooding on streams and rivers traversing the eastern Puget Lowland is common and can be severe. One of the results of floods is the development of new channels. Since the valley floors are so flat, it is relatively simple for new paths to develop as floodwaters recede. These new courses frequently remain, as in the revision of the course of the Skykomish River downstream from Monroe. The former course was westward toward Snohomish; the Skykomish River now runs southwestward, having shifted from the north to south side of Bald Hill, and joins the Snoqualmie River several miles upstream from the earlier confluence.

CHAPTER SIX

Puget Sound Bathymetry

Puget Sound is the portion of the Puget Lowland that is flooded. The processes that were described earlier shaped the Puget Lowland so that part of it was below sea level. Since the retreat of the last glaciation, the regional elevation of the lowland has increased. This was due partially to crustal changes resulting from continuing plate boundary convergence and partially to a rebounding of the earth's crust following the removal of the great weight of the ice in the Puget Lobe.

The flooded region covers an area that changes from 2,632 square kilometers (1,016 square miles) to 2,329 square kilometers (899 square miles) as the tide rises and falls. The principal connection between Puget Sound and the rest of the regional estuarine system is at Admiralty Inlet, between Middle Point on the Quimper Peninsula and Point Partridge on Whidbey Island. Two other minor connections are Deception Pass at the north end of Whidbey Island, and the Swinomish Slough, which connects Skagit Bay and Padilla Bay east of Anacortes.

Identification and description of the shape of the submarine landforms is termed *bathymetry*, and the bathymetric features of Puget Sound result in four subdivisions: the Main Basin, Whidbey Basin, Southern Basin, and Hood Canal Basin (Figure 6.1). These subdivisions are based on geographic positions, but—with the exception of the Whidbey Basin—they are also defined as basins in a bathymetric sense. The term *basin* implies a depression in the seafloor where deeper water in the middle is separated by shallower depths from deeper water beyond. In a general sense, the shallower depth separating one basin from another may be relatively insignificant, or it may be a full-fledged barrier preventing flow of water from one basin to another.

Water exchanging between Puget Sound and other portions of the regional estuarine system must flow across the shallow depths of Admiralty Inlet. Where the depth of water is shallow enough to impede free exchange of water, as it is at this main entrance to the Sound, it is termed a *sill*. It is the presence of such a sill at Admiralty Inlet that defines Puget Sound as a basin. Additional sills within the larger basin subdivide the larger system into subbasins. Oceanographically, sills are important because they impose constraints on the circulation of water deep in a basin. Water near the surface can circulate freely across the sill; but water deeper than the sill cannot move freely across it. In extreme cases, when the basin is much deeper than the sill, circulation of deep water is seriously impeded, and it may age, losing some of its life-giving oxygen. For Puget Sound, Admiralty Inlet sill is shallow enough to impede the circulation and affect some of the characteristics of the deeper water, but is not so shallow as to result in stagnant deep water.

The Main Basin

The largest subdivision of Puget Sound is the Main Basin (Figure 6.2), which covers about 45 percent of the area and holds about 60 percent of the water of the whole system. The Main Basin extends 100 kilometers (55 miles) southward from Admiralty Inlet to Commencement Bay. It is connected with the Strait of Juan de Fuca via Admiralty Inlet and each of the three other principal subdivisions of the Sound occupy peripheral basins that are connected with it. The deepest water in Puget Sound, over 280 meters (over 920 feet), is located in the Main Basin just south of the Kingston-Edmonds ferry route. The Main Basin may be considered as two parts with sharply contrasting bathymetry: Admiralty Inlet and the Central Basin.

Admiralty Inlet extends from the northern limit of Puget Sound southeastward to the southern tip of Whidbey Island. Admiralty Inlet has an average depth that is less than half that of the Central Basin. The sill in Admiralty Inlet, which is only 65 meters (215 feet) at its shallowest point, lies between the north Quimper Peninsula near Port Townsend and Whidbey Island north of Admiralty Bay. To the southeast the water deepens into a poorly defined basin east of Marrowstone Island, then shoals slightly over irregular bottom before deepening into the Central Basin southward from Point No Point and Useless Bay.

There are some contrasts between the western and eastern shore of Admiralty Inlet; the west side has higher shore cliffs, a steeper coastline, and two shallow marginal basins at Port Townsend and Kilisut Harbor. The basins are associated with the northernmost group of west channel islands, Marrowstone and Indian islands. Port Townsend occupies a shallow (20 to 25 meters, 65 to 80 meters) depression between the Quimper Peninsula and the northern portion of Indian Island. There is an extremely shallow dredged channel joining the head of Port Townsend southward to the head of Oak Bay which, in turn, opens immediately to the east and south into Admiralty Inlet. Kilisut Harbor technically separates Marrowstone and Indian islands but it is extremely shallow, almost closed off by sandbars at the north, and effectively closed off at the south by a highway that connects the two islands. Port Ludlow, about five kilometers (three miles) south of Oak Bay, also opens into the west side of Admiralty Inlet just north of Tala Point and the entrance to Hood Canal.

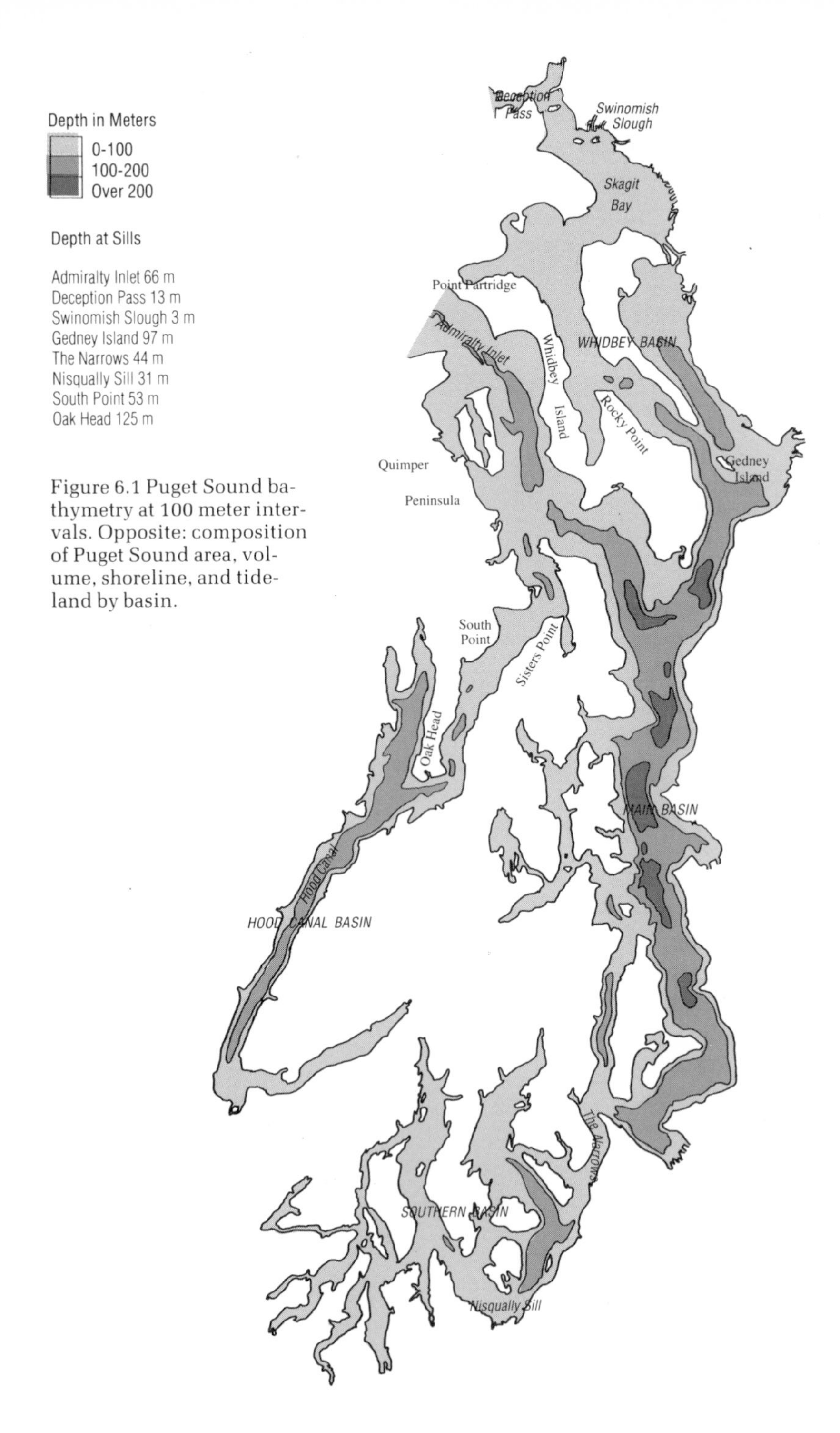

Figure 6.1 Puget Sound bathymetry at 100 meter intervals. Opposite: composition of Puget Sound area, volume, shoreline, and tideland by basin.

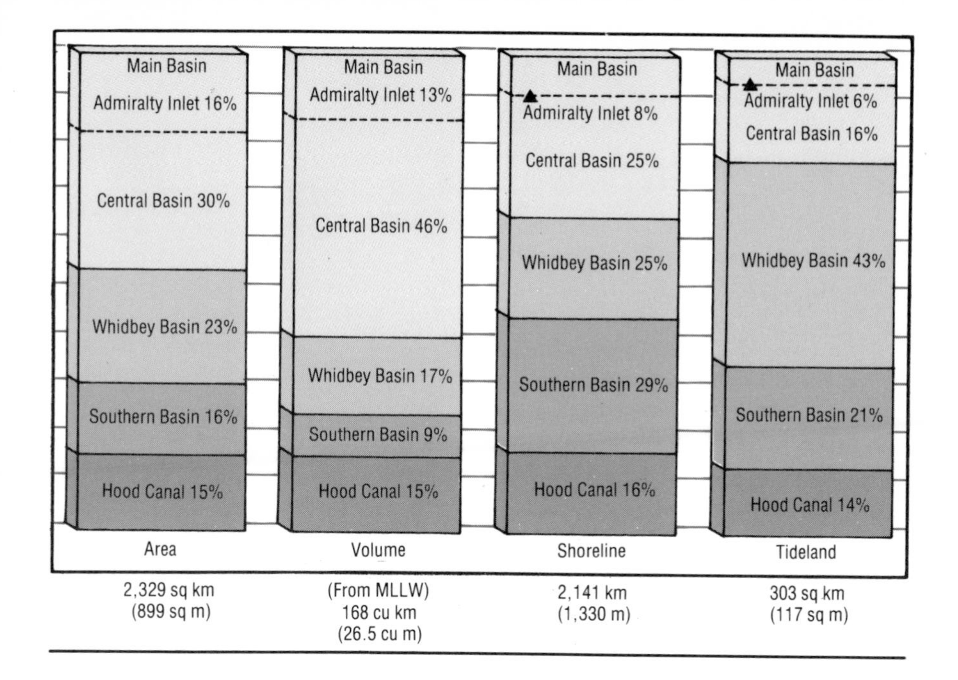

The eastern side of Admiralty Inlet is relatively straightforward. At the north, Admiralty Bay, Mutiny Bay, and Useless Bay form broad, open indentations into the western shore of Whidbey Island. Cultus Bay, a bay in name only and considerably less useful than Useless Bay, indents the south shore of Whidbey Island just north of the entrance to Possession Sound.

Although the beaches along both shores of Admiralty Inlet are generally sandy, there are rocky beaches on the Quimper Peninsula near the northern end of the inlet and some mudflats and marsh along the shore of Whidbey Island near Useless Bay and southward to Cultus Bay.

The other major subdivision of the Main Basin is the Central Basin. It extends southward from Whidbey Island to Commencement Bay. It is larger in area and volume than Admiralty Inlet and is the largest of the subdivisions of Puget Sound. At the north, it connects with Admiralty Inlet and, between Possession Point and the mainland, extends northeastward to connect with the Whidbey Basin. At the south, it connects via The Narrows with the Southern Basin.

As in the case of Admiralty Inlet, there are some prominent contrasts between the eastern and western portions of the Central Basin because of the presence of two prominent west channel islands, Bainbridge and Vashon. The western shoreline is considerably more complex than the eastern, because of the waterways that lie westward of the islands along the eastern side of the Kitsap Peninsula.

Along the western side of the Central Basin the shoreline is, with

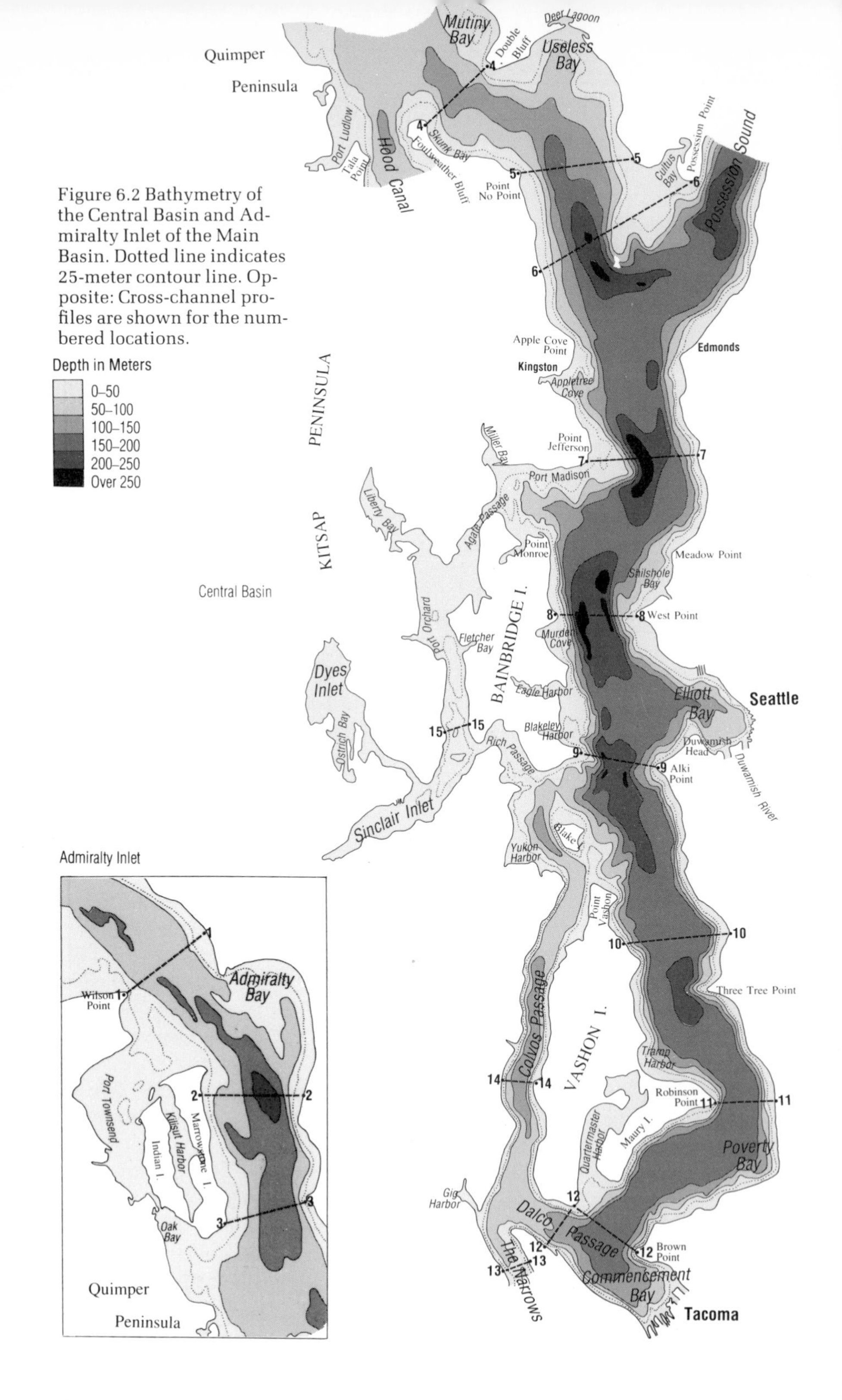

Figure 6.2 Bathymetry of the Central Basin and Admiralty Inlet of the Main Basin. Dotted line indicates 25-meter contour line. Opposite: Cross-channel profiles are shown for the numbered locations.

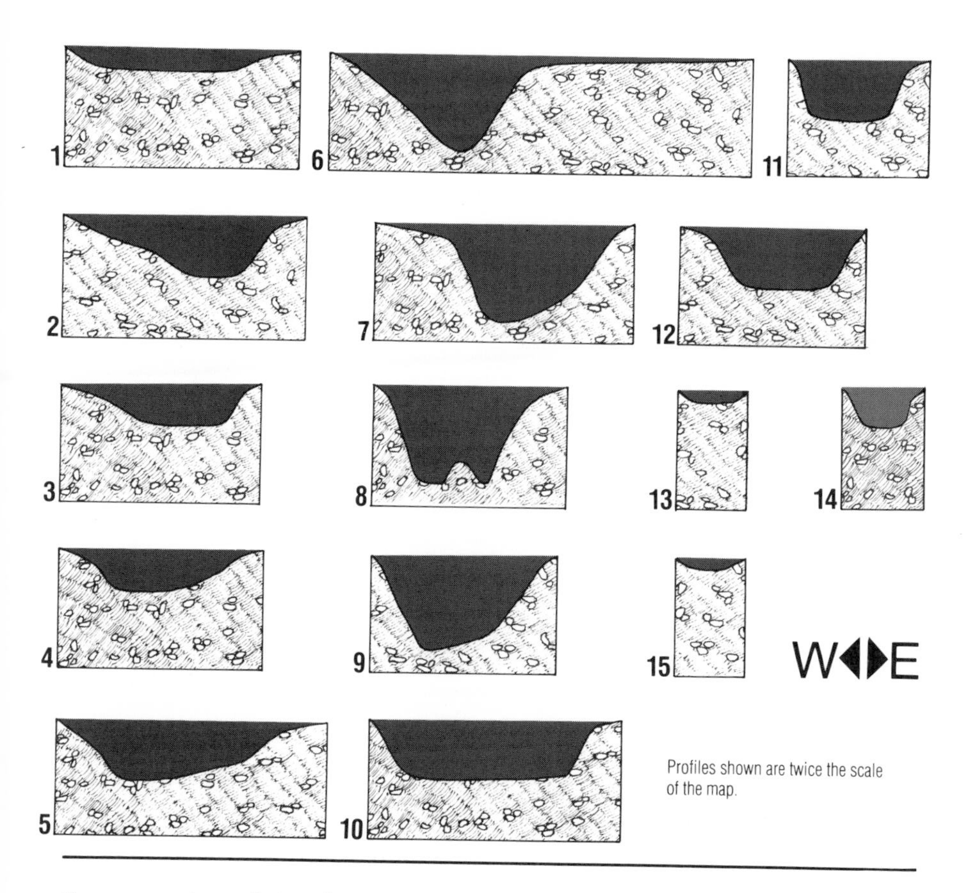

Profiles shown are twice the scale of the map.

the exception of Appletree Cove at Kingston, relatively regular southward from Foulweather Bluff until Port Madison. Here, a prominent, deep (24 to 45 meters; 80 to 150 feet) body of water opens directly into the Central Basin from the west, between the Kitsap Peninsula and the north end of Bainbridge Island.

To the west of Bainbridge Island there are many shallow inlets and channels that were formed by water from the Central Basin flooding laterally over lower portions of the glacial deposits of Kitsap Peninsula. They are called West Sound Inlets, and they include Dyes Inlet, Sinclair Inlet, Port Orchard, and Liberty Bay. At the north, they have access to the Central Basin through Agate Passage at Port Madison. At the south, Rich Passage opens into the main channel between southern Bainbridge Island and Blake Island. In contrast to the deep water in Rich Passage, Agate Passage is shallow, narrow, and difficult to traverse when the tidal currents are running strong. Although the West Sound Inlets cover about ten percent of the area of the Central Basin, they contain less than two percent of the water. As the tide rises and falls over shallow, gentle slopes of the seafloor, broad tidelands are alternately covered and uncovered. Although the West Sound Inlets cover only a small area, they account for about 27 percent of all tideland in the Central Basin, and these tidelands are predominant characteristics of the shoreline in this sector.

Farther south, Vashon and Maury islands add some complexity to the western shoreline of the Central Basin but not to the degree that Bainbridge Island and the West Sound Inlets do. Colvos Passage along the west shore of Vashon is a straight north-south waterway separating the island from Kitsap Peninsula. At the south of Vashon, Dalco Passage provides access northward to Colvos Passage, southward to The Narrows, and eastward to the southern end of the Central Basin. Extending to the north between Vashon and Maury Islands (now connected at Portage), Quartermaster Harbor is the southernmost of the several prominent basins and inlets that characterize the western shore of the Central Basin.

Compared with the west, the eastern shore of the Central Basin is by far the simpler one. There are no features similar to the West Sound Inlets west of Bainbridge and there are only two places where an otherwise regular shoreline is embayed. These are Elliott Bay at Seattle and Commencement Bay at Tacoma. They are both bays (that is, relatively deep) rather than inlets or passages; they are the sites of principal industrial and port development; and they are both sites for entry into the Central Basin of prominent rivers, the Duwamish and Puyallup.

Elliott Bay is the principal commercial harbor on Puget Sound, and it opens westward into the Central Basin between West Point and Alki Point. The water depth over much of the Bay is in excess of 100 meters (325 feet), and depths in excess of 25 meters (80 feet) occur at many places nearshore. In contrast with the West Sound Inlets, Elliott Bay is a widening of the deeper portion of the main channel. The southern shore of the bay has been highly altered by industrial and port development but still provides access to the Duwamish-Green River system.

The portion of the Central Basin southward from Elliott Bay is called East Passage because it is the eastern passage around Vashon Island. There is, however, no bathymetric or oceanographic reason to consider East Passage a subdivision of the Central Basin since it is primarily the southward extension of the main channel.

Commencement Bay is at the southern end of the Central Basin. Much of the commercial development of Tacoma is located on the extensive delta of the Puyallup River but the water in the Bay deepens rapidly to the northwest and joins the deeper portion of the main channel southwest of Brown's Point.

Most of the shoreline of the Central Basin is bordered by low cliffs, hilly terrain displaying moderate relief and gentle slopes. Beaches are primarily sandy but some rocky areas are found, for example, at Tahlequah and Point Vashon on Vashon Island.

A juxtaposition of submarine and onshore topographic conditions has had a major impact on commercial development of the local region. Major ports have two principal requirements: ships must have deep water close to shore, and there must be access to the port's hinterland. These conditions are best combined on the eastern shore of the Central Basin. The deep water is close to shore and the shoreline indentations at Elliott Bay and Commencement Bay provide shelter for anchorage and docking. Further, these two bays are at the mouths of broad, glacial and alluvial-filled, flat-bottomed rivers that provide access routes to the shore for railroads that serve the region.

The Whidbey Basin

The Whidbey Basin contains the waters eastward of Whidbey Island and covers the second largest area of the four subdivisions of Puget Sound (Figure 6.3). Whidbey Basin is more properly a geographic term than one justified by oceanographic or bathymetric criteria. There is no obvious sill across the entrance, which is somewhat arbitrarily established at an imaginary line between Possession Point on Whidbey Island and Meadowdale on the mainland. Because of bathymetric continuity of deeper water from the northern end of the Central Basin into Possession Sound, Whidbey Basin could be considered an appendage to the Main Basin.

From the deeper water at the southern limit in Possession Sound the depth of water in the Whidbey Basin generally shoals northward. Whidbey Island forms the western shoreline of the Whidbey Basin; the eastern shoreline is the mainland. Geographically, the basin has four subdivisions because of the presence of Camano Island in the center of the area: Skagit Bay at the north, Port Susan at the east, Possession Sound at the south, and Saratoga Passage at the west.

Skagit Bay is the shallowest of the Whidbey Basin subdivisions with a mean depth of only 8 meters (25 feet). From the shallow east end, it deepens westward toward Whidbey Island. This assymetrical cross-distribution of water depth is a result of shoaling along the eastern side where the delta of the Skagit River is encroaching upon the bay. The Skagit is the largest river in the Puget Sound system, and is depositing sediments into the bay that have created extensive tidelands and mudflats. Viewed from land or water, these tidelands are the most prominent characteristic of Skagit Bay and are unequalled anywhere else in the Puget Sound system.

Geographically, Skagit Bay has a natural outlet with the eastern end of the Strait of Juan de Fuca via Deception Pass, and with Padilla

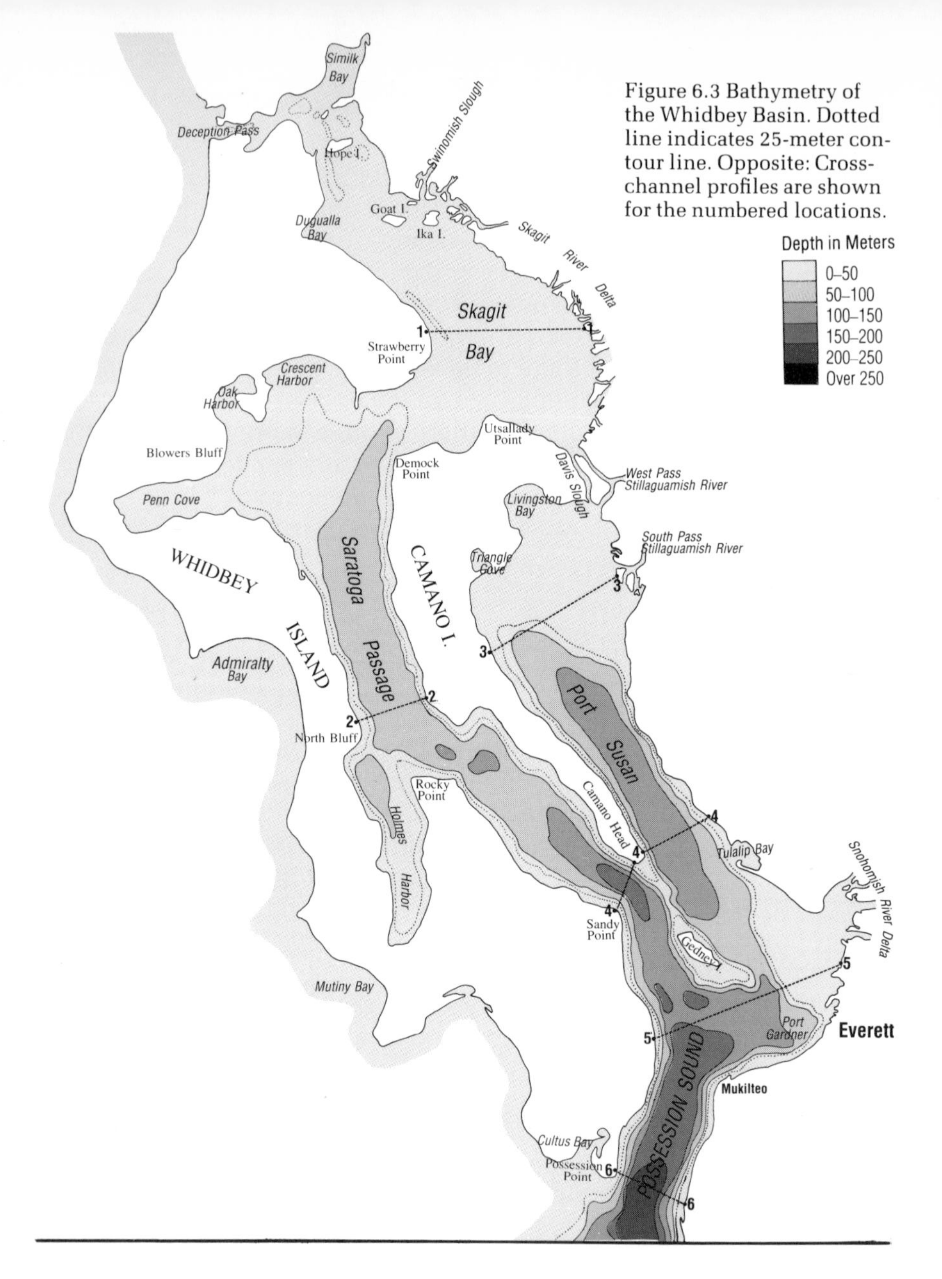

Figure 6.3 Bathymetry of the Whidbey Basin. Dotted line indicates 25-meter contour line. Opposite: Cross-channel profiles are shown for the numbered locations.

Bay via the Swinomish Slough. Of these two northern outlets of Puget Sound, Swinomish Slough is the less spectacular. It traverses the low-lying land connecting Fidalgo Island to the mainland. Because of the configuration of the Whidbey Basin, the level of the tides inside Skagit Bay are frequently out of phase with those outside and there can be a resultant strong current flow through Swinomish Slough. For similar reasons, the tidal current is even more obvious through Deception Pass, which connects westward from Skagit Bay between Fidalgo and Whidbey islands. Although Deception Pass has been called picturesque be-

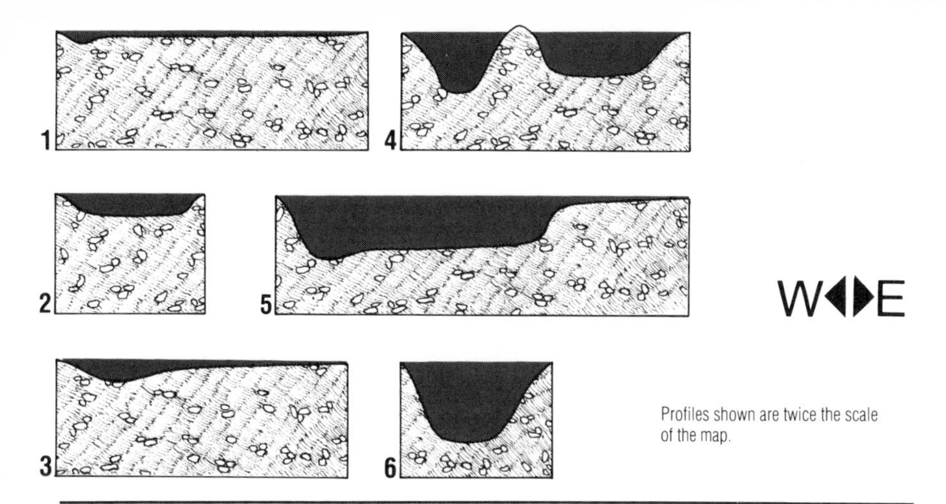

Profiles shown are twice the scale of the map.

cause of its rocky and narrow topography, when strong currents are running, it provides a very rough run for small boats.

At the south, Skagit Bay connects with the northern end of Saratoga Passage between Strawberry Point (Whidbey Island) and Utsalladdy Point (Camano Island). Camano Island was once separated from the mainland, but the connecting waterway between Skagit Bay and Port Susan has been filled by sediments deposited by the encroaching Skagit River delta from the north and the Stillaguamish River west of Stanwood from the south.

Saratoga Passage contains deep water separating Whidbey and Camano Islands. Opening southward from the shallow waters of Skagit Bay, the main channel in Saratoga Passage deepens to 177 meters (580 feet) between Sandy Point and Camano Head near its southern end. Three shallow areas—Penn Cove, Oak Harbor, and Crescent Harbor—open westward from Saratoga Passage and combine to form the widest, east-west, portion of the Passage at its northern end. At the south, Holmes Harbor is a large, deep indentation into the eastern shore of Whidbey Island. At its mouth, the northern end of Holmes Harbor is separated from the main channel of Saratoga Passage by a small spit formed by bay mouth deposits resulting from material that has been moved along Whidbey Island's western shore by currents. At its head, Holmes Harbor is now separated from Mutiny Bay on Admiralty Inlet by about a mile-wide portion of low-lying land that was probably formed quite recently in postglacial times. Southward from the mouth of Holmes Harbor, the main channel of Saratoga Passage shoals slightly west of Gedney Island before deepening southward onto Possession Sound.

Port Susan contains the deeper water of the Whidbey Basin between Camano and Gedney islands and the mainland. The northern end is relatively shallow and is being filled by deltaic deposits of the

Stillaguamish River. These deposits have shoaled an earlier connection northward with Skagit Bay and have also resulted in extensive shoals and mudflats, which almost fill areas such as Livingstone Bay and Triangle Cove on Camano Island. Deepening southward, the main channel in Port Susan reaches its greatest depth (123 meters; 400 feet) eastward of Camano Head at the southern tip of the island. Farther south the Port Susan sill is a subcritical sill with a controlling depth of 97 meters (320 feet) and of relatively recent origin. The submarine ridge extending southward between Camano Head and Gedney Island forms the western boundary of the deeper water in Port Susan and the main channel lies parallel and to the east of the ridge. Sediments being deposited by the Snohomish River are encroaching westward across the main channel and the constriction gives rise to Port Susan sill.

Possession Sound lies between the southern ends of Saratoga Passage and Port Susan, and the deeper water of the Main Basin of Puget Sound. Bathymetrically it is more properly a portion of the Main Basin since there is a continuity of the deeper channel of the Main Basin northeastward into Possession Sound.

Overall, the Whidbey Basin shoreline is a contrast between the islands and the mainland. Most of the insular shoreline is backed by low cliffs and hilly terrain, as is part of the mainland. On the other hand, the extensive mudflats and marshes that characterize the river deltas on the mainland are backed by very flat terrain. Beaches on Whidbey and Camano islands and along the mainland shore of Port Susan and Possession Sound are sandy in contrast with the mudflats associated with the major river deltas.

The Southern Basin

Opening southward from The Narrows, the Southern Basin (Figure 6.4) of Puget Sound is the historic "Puget's Sound." The area of the Southern Basin—449 square kilometers (173 square miles) at mean high water, about 15 percent of which is tidelands—ranks only third among the four principal subdivisions of Puget Sound. The region was originally identified by Captain George Vancouver (c 1792), but detailed exploration and mapping of its many miles of inlets, passages, islands, and interconnected waterways was carried out by his lieutenant, Peter Puget. The true scope of Puget's accomplishment can be best appreciated by a comparison of the shoreline complexity of the Southern Basin with the rest of the Puget Sound system. The Southern Basin has significantly more shoreline than any other portion of Puget Sound. Much of this complexity is due to the presence of several prominent islands (Fox, McNeil, Anderson, Hartstene) and an assemblage of shallow inlets (Hammersley, Totten, Eld, Budd, and Henderson).

The entrance sill of the Southern Basin is in The Narrows, a short, steep-sided passage that provides the only access between the Southern Basin and the Main Basin. The controlling sill depth (the deepest connection over the sill) is just under 45 meters (150 feet) and the main channel deepens rapidly southward to a maximum depth of 188 meters (615 feet) east of Balch Pass. In spite of this maximum depth, most of the Southern Basin is characterized by shallower waters; the mean depth of the whole of the Southern Basin is only 37 meters (120 feet), significantly less than any of the other principle subdivisions of Puget Sound.

Bathymetrically the Southern Basin is deepest along a curving channel extending from between McNeil and Fox islands southward, then westward around Anderson Island, and northwestward into southern Case Inlet. At its southernmost extremity, this channel is being impinged upon from the south by the outbuilding delta of the Nisqually River onto Nisqually Flat. In a manner similar to the Snohomish River delta in Port Susan, the channel southward from Anderson Island has developed a midchannel sill which effectively separates the bottom waters in the east from those to the west. The controlling depth at the Nisqually Sill is only 31 meters (100 feet) and the channel deepens northwestward to a basin depth of about 110 meters (360 feet) midchannel between Johnson Point and Devils Head.

Of the several inlets in the Southern Basin the most prominent is Carr Inlet. From its north end in the shallow and now cut-off Burley Lagoon, the channel deepens southward through Henderson Bay and contains a large area where depths are greater than 150 meters (490 feet). The mean depth of Carr Inlet is 50 meters (165 feet)—relatively deep—indicating some degree of glacial scouring of the principal channel. The presence of Fox and McNeil islands adds to the complexity of the shoreline at the southern end of Carr Inlet. Pitt Passage is a shoal separation between McNeil Island and the southern extension of the Kitsap Peninsula. Hale Passage, somewhat deeper, separates Fox Island from the mainland. Each of these passages provides an interconnection into southern Carr Inlet.

At first glance, Case Inlet appears to be a twin to Carr Inlet, but the relationship is fraternal rather than identical. Of comparable overall length and general north-south orientation, Case Inlet is not as deep as Carr and the main channel is somewhat more complex. From the shallow waters of Vaughn, Rocky, and North bays at its northern end, the main channel deepens southward from the entry of the north end of Pickering Passage. The channel is broader and has more gently sloping lateral gradients than Carr Inlet. Channel depth in excess of 40 meters (130 feet) extends southward through a minor interior basin located southwest of Herron Island, over a slightly more shoal area, and into the deepest water of the western part of the Southern Basin.

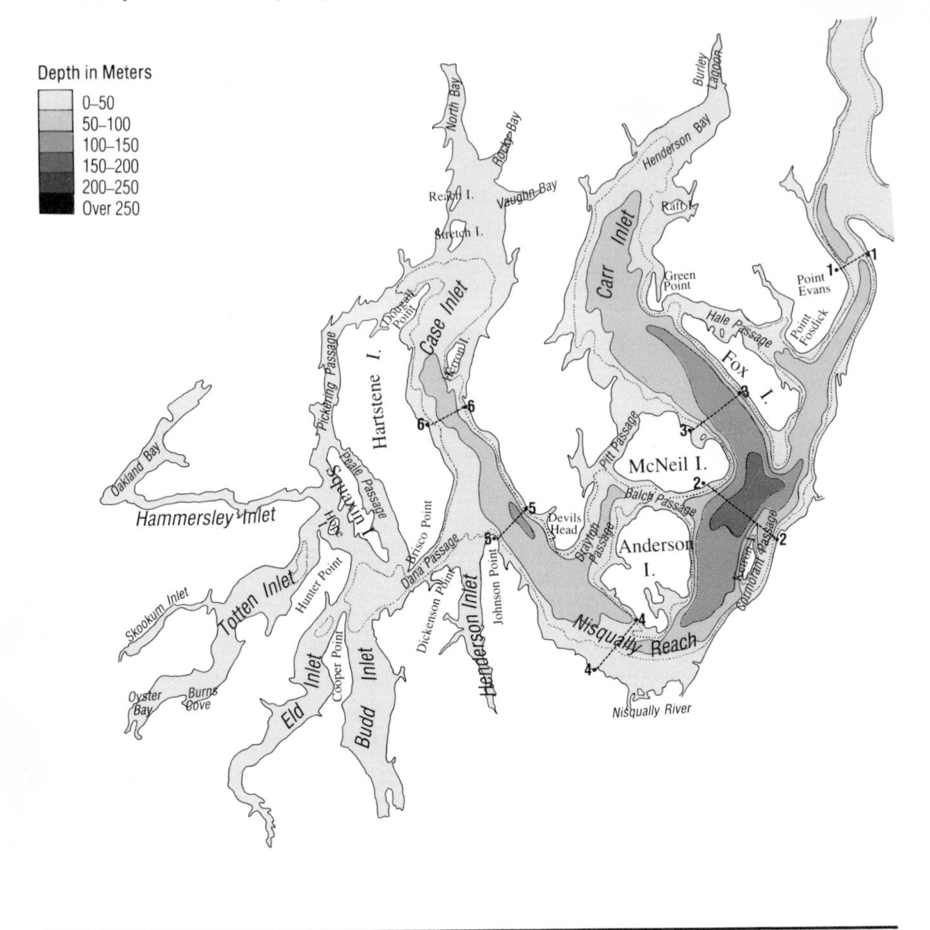

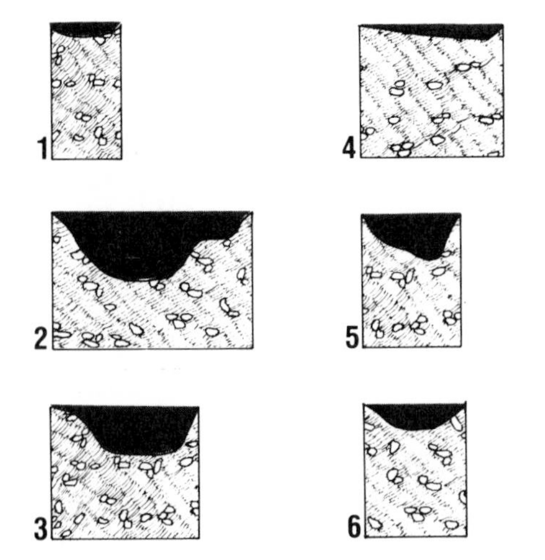

Figure 6.4 Bathymetry of the Southern Basin. Dotted line indicates 25-meter contour line. Cross-channel profiles are shown for the numbered locations.

Profiles shown are twice the scale of the map.

A very distinctive portion of the western part of the Southern Basin lies south and west of Hartstene Island and is generally the shallowest portion of the entire Puget Sound system. This region is characterized by passages (Pickering, Peale, Dana, Squaxin) and inlets (Hammersley and Oakland Bay, Totten and Skookum, Eld, Budd, and Henderson). The passages are open at both ends and separate the two major islands (Hartstene and Squaxin) from the mainland. In response to the Southern Basin's large tidal range, the passages are frequently subject to strong tidal currents. In contrast, the inlets are open at only one end and are characterized by large intertidal areas resulting from the combination of large tidal range and gentle slopes associated with inlets. The ratio of tideland to high-water area in this portion of the Southern Basin is comparable to that in the Whidbey Basin. But here the tidelands are not the result of active sedimentation by the region's major river deltas: Southern Basin streams are of secondary importance to the combination of tide range and low gradients. In contrast to Whidbey Basin, Southern Basin tidelands tend to be localized at the heads of inlets.

As over much of Puget Sound, shorelines in the Southern Basin are backed by hilly terrain with gentle slopes. Sandy beaches are common but rocky areas occur in a number of localities, perhaps more commonly in the passages and inlets at the southwest. Mudflats are common along the Nisqually River delta, in the shallow headwaters at the north ends of Case and Carr inlets, and extensively in the heads of the small inlets at the southwest.

The Hood Canal Basin

Hood Canal Basin is the smallest and least complex of Puget Sound's subdivisions (Figure 6.5). Geographically, the entrance to Hood Canal opens westward from Admiralty Inlet between Tala Point and Foulweather Bluff. The entrance sill is farther west where a sill with a controlling depth of just over 50 meters is located between South Point and Lofall.

The portion of the Hood Canal system between Foulweather Bluff and the southern tip of Toandos Peninsula is termed the Hood Canal entrance. The major axis of deeper water in the Hood Canal Basin extends from Dabob Bay southward to the region of Potlatch and Annas Bay. The Hood Canal entrance joins this deep, linear feature from the northeast about one-third of the distance between Dabob and Annas bays. This juncture, between Oak Head and Seabeck Bay, is marked by a minor shoaling but not by a controlling sill.

The main axis of the deep channel of Hood Canal occupies a glacially scoured depression along the western boundary of the Puget Lowland. In contrast with the Main Basin of Puget Sound, which was eroded into the middle portion of the Puget Lowland, Hood Canal was

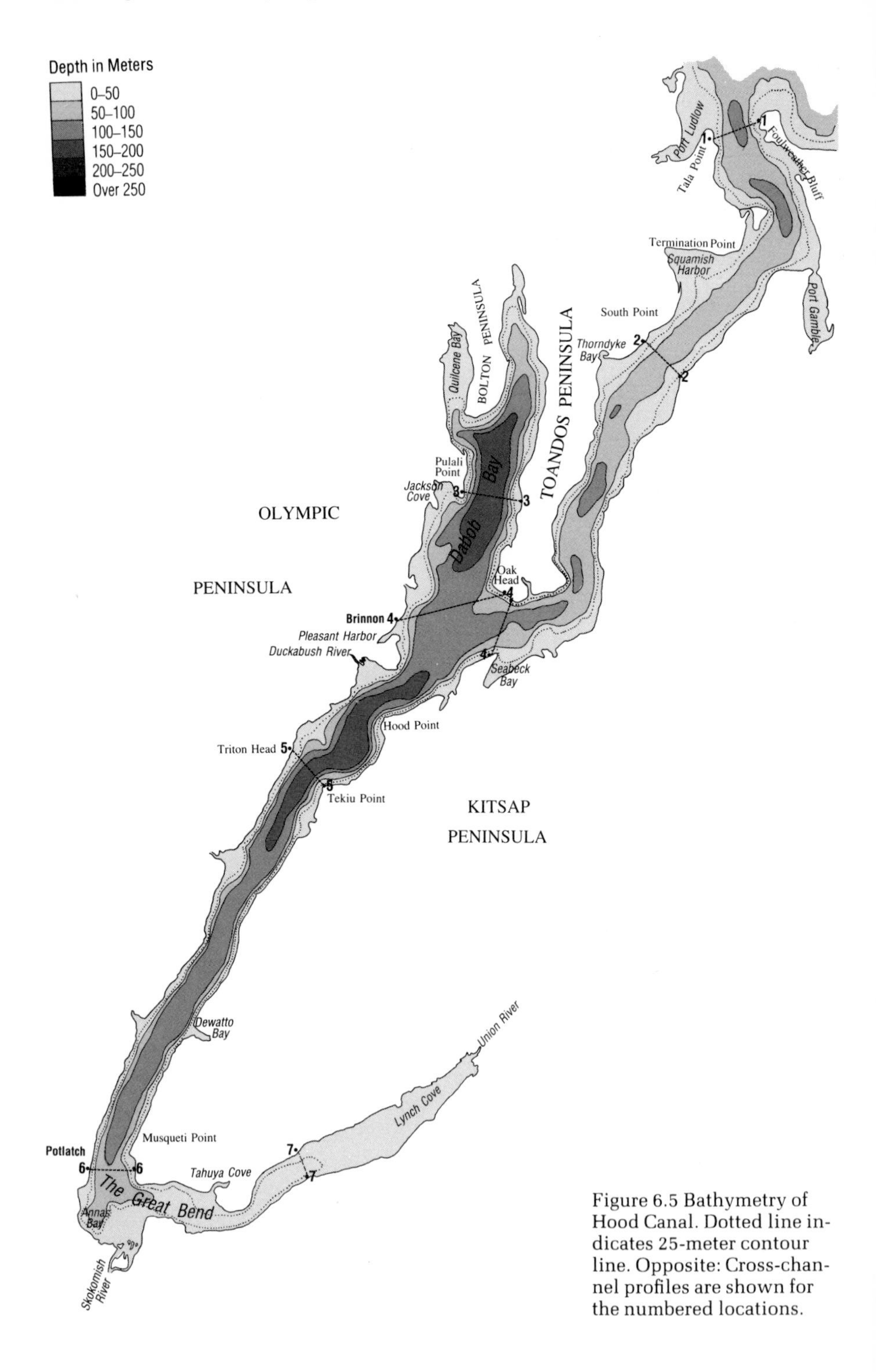

Figure 6.5 Bathymetry of Hood Canal. Dotted line indicates 25-meter contour line. Opposite: Cross-channel profiles are shown for the numbered locations.

W◀▶E

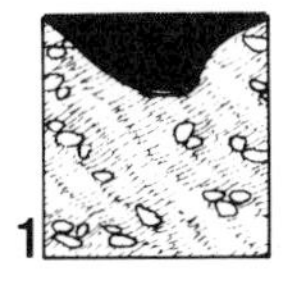

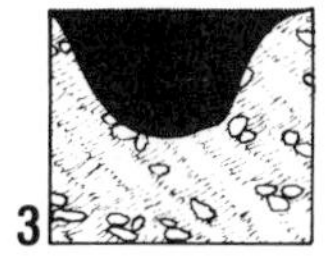

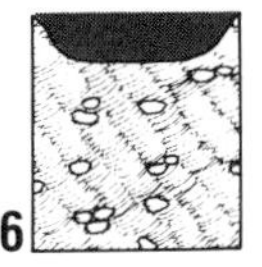

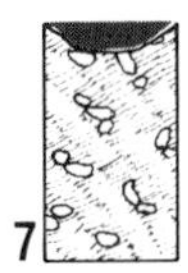

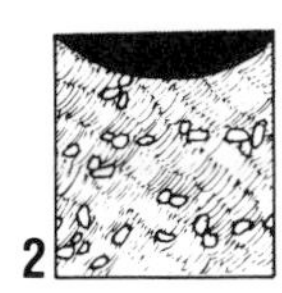

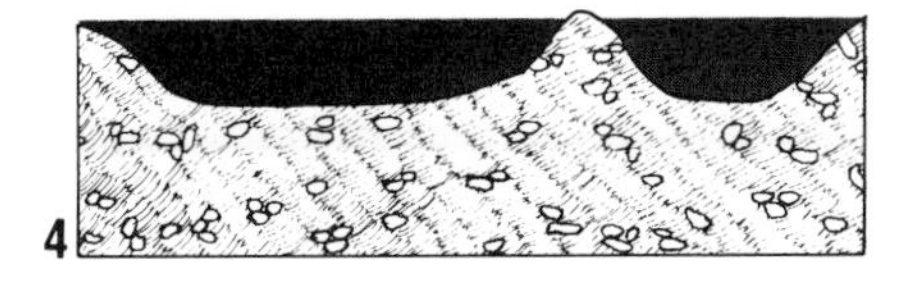

Profiles shown are twice the scale of the map.

---

eroded along the western margin of the Puget Lobe. This scoured channel does not include the Hood Canal entrance, but extends from Dabob Bay southward through the main channel of Hood Canal to Annas Bay. The deepest water, over 185 meters (600 feet), is in Dabob Bay between Pulali Point and the Toandos Peninsula. Dabob Bay is separated from the main channel of Hood Canal by an intermediate sill west of Oak Head. Water deepens southward into the main channel reaching midchannel depths of more than 180 meters (590 feet) west of Hood Point and Tekiu Point. Farther south the maximum midchannel depth shoals uniformly and ends on the sand and mudflats that mark the entrance of the Skokomish River into Annas Bay.

The south end of Hood Canal is termed "The Great Bend." At sea level, the bend appears to be a prominent turning of the axis of Hood Canal where it hooks to the northeast. Bathymetrically, the axis of Hood Canal does not actually turn: the water eastward from the Great Bend is a relatively simple appendage that joins the main system at its southernmost extremity. To the east of the Great Bend there is a general shoaling northeastward into the mudflats of Lynch Cove at the mouth of the Union River. In contrast with the evidence of glacial scouring associated with the main axis of Hood Canal, the waters between the Great Bend and Lynch Cove are shallow and occupy part of a former glacial meltwater distributary that extended across the Kitsap Peninsula from Lynch Cove to Sinclair Inlet.

In contrast to all of the other subdivisions of the Puget Sound System, Hood Canal has very limited tidelands, marginal bays, coves, and mudflats. In the Hood Canal entrance, Squamish Harbor and Thorndyke Bay are only minor indentations of the western shoreline and Port Gamble, although larger, is extremely shallow and almost isolated by shoal water near its mouth. Along the main axis of the system, Quilcene

Bay is a prominent embayment opening into the northern end of Dabob Bay from the west at the southern tip of the Bolton Peninsula. At the southern end, Annas Bay reflects shoaling by sedimentation on the delta of the Skokomish River. The extensive mudflats at the head of Lynch Cove are, in contrast, tidelands formed in a manner similar to those in the inlets of the Southern Basin.

Although the shoreline of the Hood Canal Basin is the shortest and least complex of the major subdivisions of Puget Sound, the western shore along the main axis is backed by rugged coastal topography. To the west, high cliffs are backed by mountainous terrain of the eastern Olympic Mountains, which are very close to the shore west of Dabob Bay. The eastern shoreline is backed by the more subdued hilly terrain of the Kitsap Peninsula. Beaches are predominantly sandy with barrier islands across the mouths of smaller inlets (e.g., Pleasant Harbor near Brinnon). Muddy beaches and mudflats are associated with deltas of some of the local rivers (e.g., Duckabush, Skokomish) and at the heads of larger bays and inlets (e.g., Quilcene, Lynch Cove).

CHAPTER SEVEN

The Changing Shape of Puget Sound

The volume of the Puget Lowland below sea level was greatest when the Puget Lobe glacier occupied the region. The powerful moving ice cut deep channels into the underlying land, and its tremendous weight caused a regional downwarping of the earth's crust. Consequently, although glacial ice rather than seawater filled the lowland, the land surface below the ice was generally well below sea level. If some catastrophic process could have removed the ice almost instantaneously and permitted the marine waters to fill the depression, Puget Sound would have been deeper and covered more area than it does today. What we observe today is a smaller feature which, from the large volume of the glacial age, has been growing smaller over the years and is continuing to shrink.

From the time of maximum size, about thirteen thousand years ago, two processes have been at work reducing the volume of Puget Sound. The first process is associated with the withdrawal of the Puget Lobe. Once the Vashon glacial ice retreated and the tremendous weight of the ice was removed the level of the land surface rose. Evidence of this emergence of land can be seen in elevated beaches found, for example, near Bellingham. These are features originally formed at shoreline when sea level was higher than it is today. The melting of the glacial ice also resulted in an increase of total volume of water in the oceans and a net raising of sea level. The combination of the relative ups and downs of both land surface and sea level has been complex in the region, and the cumulative result has been a decrease in the volume of Puget Sound.

The second process associated with retreat of glacial ice is perhaps more easily observed; it involves filling in of depressions with sediment. Sedimentation, a universal geologic process, involves movement of material from uplands and mountains downhill to the lowlands and ultimately into bodies of marine water and the oceans. Sedimentary processes work in one direction—from above sea level to below sea level. They are at work continuously across the surface of the earth and are responsible for slowly modifying both subaerial and submarine forms worldwide. Sedimentary materials are deposited into Puget Sound from the Cascades, the Olympics, and the Puget Lowland. In the

Sediment Type	Grade Limits
BOULDERS	Greater than 256mm
GRAVEL	
Cobbles	256-64 mm
Pebbles	64-4 mm
Granules	4-2 mm
SAND	
Very coarse	2-1 mm
Coarse	1½ mm
Medium	1/2-1/4 mm
Fine	1/4-1/8 mm
Very fine	1/8-1/16 mm
MUD	
Silt	1/16-1/256 mm
Clay	Less than 1/256 mm

Table 7.1 Classification of sediments by size

absence of renewal of uplands by crustal mountain building processes, the sedimentary process could ultimately wear down higher landforms and fill in much of Puget Sound. In the following pages we will examine the characteristics of sedimentation in the local area with particular attention to how it is changing the shape of Puget Sound.

The Sedimentary Process

Most rocks resulting from geologic processes are generally too massive to be moved along the earth's surface by anything short of a major natural catastrophe. In contrast, particles moved by sedimentary processes are small, and originate in the almost universal cover of unconsolidated fragments of broken rocks and organic plant debris. The thickness of this unconsolidated layer, the size and shape of the individual fragments, and the mix of organic and inorganic material vary from place to place. Fragment size can range from large boulders to particles too fine to be observed by the naked eye. Although there are many specific categories of particles based on size (see Table 7.1), use of the common names (boulders, gravel, sand, and mud) simplifies discussion of erosion, transport, and deposition processes.

The edges of rock fragments can range from extremely angular to smoothly rounded, and the shape can be elongated, flattened, or spherical. Fragment assemblages may consist of similar-sized particles (well-sorted) or a diverse mixture of particle sizes (poorly sorted). The distribution of specific particle sizes, shapes, and sorting is caused by composition and characteristics of the parent rock material, the processes that break up the parent rock, the processes that move the fragments, and conditions at the site of deposition. Study of these fragments, individually and in aggregate, provides geologists with insight into where the sedimentary particles came from, how they were moved, and conditions under which they accumulated.

Weathering

When rocks formed within the earth's crust become exposed at the earth's surface, they are thrust into a different environment. They are subject to different temperature and pressure at the surface and (of major importance) become exposed to water, one of the most universally active natural chemical agents or solvents.

The general response to this disequilibrium is a disintegration of the original rock into smaller fragments. In a purely disintegrative breakup of rock, termed *physical weathering*, there is no change in mineral composition but simply the creation of small particles from big ones. Yet weathering can also involve chemical and mineralogical alteration of rock as well as removal of material in solution. Rocks in the same region may weather differently depending on climatic conditions. The rate and the products of weathering depend on the following:

composition of the material;
amount of moisture (from rainfall);
temperature, particularly the extremes of hot and cold and daily and seasonal variability;
biologic activity, primarily plant growth;
topographic relief (slope) of the surface.

Another factor determining the degree of weathering is the amount of time weathering processes have been working. In the Pacific Northwest, this time has differed depending upon whether the surface was covered by glacial ice. Ongoing subaerial weathering processes have been working longer in unglaciated regions, while glaciated area had new material exposed only after the ice retreated.

Soils

When soils are young they tend to reveal the character of parent materials, but more mature soils frequently are characterized by vertical zonation that reflects climatic conditions rather than parent rock. In the Puget Sound drainage basin, three dominant soil types (Figure 7.1) contribute sedimentary material to be deposited in the Sound.

On the high western slopes of the Cascades and on the Olympic Peninsula, soils display characteristics of volcanic parent rocks that have weathered under conditions of abundant precipitation and cool temperatures of the higher altitudes. These soils are relatively young and vertical zonation is only partially developed. Although the upper layers have been leached of iron and lime, there is no well-developed

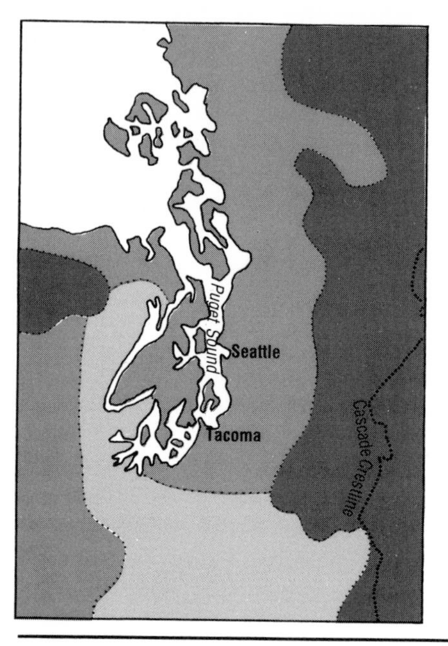

High altitude soil Brown forest soil
Well-developed Brown forest soil

Figure 7.1 Distribution of soils in the Puget Lowland east to the Cascade ridge.

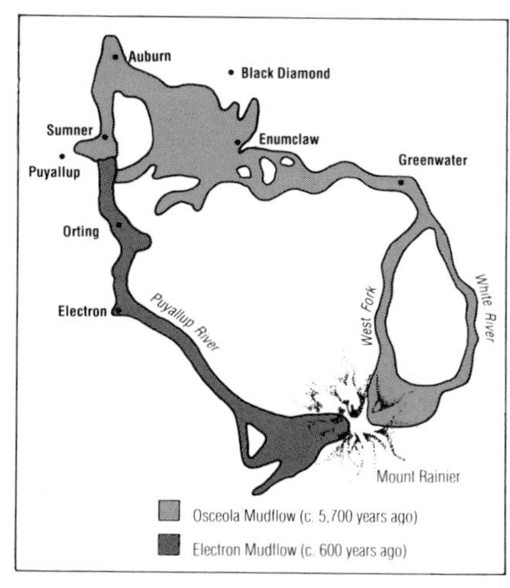

Figure 7.2 In two separate occurrences, volumes of water-saturated earth have slumped and slid off the flanks of Mount Rainier, following river valleys down into the lowland.

lower zone of accumulated leached materials.

Over the Puget Lowland, the Quimper Peninsula, and the foothills of the western Cascades there is a second soil type, a brown forest soil that supports abundant woodlands and ranges. At the surface, this soil has a thick, dark-colored layer containing a very high concentration of organic debris. Other characteristics of a well-developed vertical zonation are absent. There is only minor indication of downward leaching of material and no evidence of accumulation of leached material in a lower zone. This soil is most directly related to weathering of glacial deposits; hence it is relatively young soil, which has been developing over the last 10 to 12 thousand years. It displays a high degree of inhomogeneity, a reflection of the many localized differences in the specific type of glacial deposits. Associated with this brown forest soil are several localized occurrences of peat and marsh soils, reflecting the diversity of postglacial drainage conditions.

The third soil type is found principally in the southern portion of the Puget Lowland, which was not directly affected by glaciation. Relative to the brown forest soil, these soils display evidence of greater maturity. Their upper zone has high organic content, but has been leached of clay minerals, iron, and aluminum. Although generally not well developed, there is also a lower zone of accumulated leached minerals. These soils are subject to the same degree of high moisture and seasonally cool weathering as the brown forest soil; however, they differ in substance (nonglacial versus glacial deposits), and probably have weathered longer.

Collectively, the soils of the Puget Sound region are relatively immature, are generally acidic, support woodlands and ranges, but are not naturally good farmlands (although precipitation is adequate). These soil characteristics have resulted in land uses quite different than those east of the Cascades. The lumber industry of western Washington is founded on the abundant forests supported by the dominant soil types. Most agricultural land supports pasturage and feed for livestock (principally dairy cattle), which reflects less favorable local soils. Some of the local soil types, particularly in drained marshes and the lower Skagit River Valley, support vegetable and truck crops; but these are more likely the result of proximity to large urban centers than a natural outgrowth of favorable soil conditions.

Erosion and Transportation

The movement of soil and other unconsolidated fragments is accomplished by a number of natural processes. The role of glacial ice has already been discussed and the remaining glaciers in the Cascades and the Olympics still contribute to particle movement. Movement by wind is an important transportation process in some parts of the world, but because of heavy plant cover and high moisture, it is not significant in western Washington. In the higher parts of the uplands, downhill slumping and sliding of unconsolidated material is an important mode of migration. The most spectacular of these movements is associated with landslides, debris-avalanches, and mudflows caused by gravity acting on a combination of an oversteepened slope and water-saturated detritus. Sudden, severe downslope movement, such as the large mudflows that came off the flanks of Mount Rainier (Figure 7.2) can be triggered by seismic or volcanic activity. Less spectacular, but probably of greater overall importance, is the slow downslope creep of unconsolidated surface material over the western flanks of the Cascades, in the Olympics, and on the steeper slopes in the Puget Lowland.

Of all of the processes that transport particles, the most obvious and generally the most important is running water. The network of streams and rivers that drain the watershed of Puget Sound constitute a primary means for moving unconsolidated material.

The Pacific Northwest estuarine system is a transition area in which the freshwater inflow from rivers mixes through estuarine into true marine waters of the Pacific Ocean. For the entire regional estuarine system the dominant source of freshwater is Canada's Fraser River, which has a large drainage basin east of the Cascades. Puget Sound itself receives drainage from a relatively small area of just over 28,500 square kilometers (11,000 square miles). The total freshwater contribution of the rivers entering Puget Sound is about one-third that of the Fraser River and considerably less than the total freshwater drainage into the Straits of Georgia and Juan de Fuca.

Climate

Flow and runoff within a drainage system is strongly influenced by regional climate. In the most general sense, Puget Sound's climate results from cool, moist maritime polar air moving east by northeast off the northeast Pacific Ocean. This results in the cloudiness and the impression that the sun never really shines. An average of 150 days of the year have precipitation of more than 0.01 inch and, although this precipitation is distributed throughout the year, it is highest in winter. Although total annual precipitation is not high, low evaporation rates (due to a high degree of cloud cover and low annual incoming solar radiation) result in a damp, humid climate. The annual temperature range is relatively small when compared with inland midlatitude temperatures.

There are climatic contrasts that result from topographic influences of the Olympic and Cascade mountain systems. The Cascades are the major precipitation divide in the Pacific Northwest, with annual precipitation high to the west and low to the east. Mean annual precipitation is highest on the western slopes of the Olympics (128 inches per year), least in the Puget Lowland (40–80 inches per year), increasing up the western slopes of the Cascades to about 80–100 inches per year in the high Cascades. Throughout the year, about 50 percent of the regional precipitation falls on the western slopes of the Olympics, about 40 percent falls on the foothills and western slopes of the Cascades, and only 10 percent falls on the Puget Lowland.

Forty percent of the regional annual rainfall occurs during winter, when the flow of marine air is characterized by frequent cyclonic storms. Winter precipitation in the lowlands is principally rain, but snow is prevalent in the Cascades where annual accumulations range from 300 to 1,000 inches. The prevalent cloud cover keeps winter temperatures moderate in the lowland, although the higher mountain areas are below freezing. Also during winter, the combination of shorter days, cloud cover, and low-level incident sunshine combine to give as little as two hours a day average sunshine.

Spring is an extension of winter with decreasing frequency of cyclonic storms, some decrease in the overall precipitation, and intermittent brief absences of cloud cover. In the Puget Lowland the average temperature range is between 40° and 60°F, but in the mountains it remains cold enough to maintain snowfall until late into the spring.

Summer is statistically the best season of the year. Summertime averages account for only 10 percent of the total annual rainfall, and a seasonal modification of the general onshore windflow gives rise to less than average cloud cover over the lowland. Temperatures are at their annual highs although still relatively low in contrast with nonmaritime midlatitude regions.

Summertime conditions frequently extend into autumn but autumn is normally a period of increasing precipitation and cyclonic storms.

The Drainage System

Seasonal cycles of precipitation and temperature directly influence the characteristics of freshwater runoff from the region's rivers. In turn, these spatial and seasonal contrasts are a major factor in determining the rate and characteristics of sediment being carried into Puget Sound (Figures 7.3 through 7.6).

The Whidbey Basin

The Whidbey Basin receives over 60 percent of the total freshwater discharged into Puget Sound. The three largest rivers, Skagit, Snohomish, and Stillaguamish, enter the Whidbey Basin and drain an area covering over 14,500 square kilometers (5,600 square miles), about 50 percent of the total drainage area of Puget Sound.

The Skagit River is the largest river entering Whidbey Basin; it begins in Canada and drains much of the Northern Cascades. Major impounding of the Skagit occurs at Ross Dam (forming Ross Lake) and Diablo Dam. Downstream of Ross Lake, the Skagit is joined by the Cascade River near Marblemount, by its major tributary the Sauk River near Rockport, the Baker River south of Baker Dam, and Lake Shannon near Concrete. The lower reaches of the Skagit flow over the glacial deposits of the Skagit Valley. Although the river itself is relatively large, it occupies an oversized valley originally eroded by glacial ice rather than the river itself. It enters Skagit Bay southwest of Mount Vernon where its riverborne sediments are deposited to form extensive mudflats that occupy much of Skagit Bay.

The Snohomish River is the second largest river entering the Puget Sound system. The Snohomish system is the result of confluence of two major tributaries—the Skyhomish and Snoqualmie rivers. They join south-southwest of Monroe, forming the Snohomish River, which is joined by the Pilchuck River near Snohomish and enters Possession

Sound north of Everett.

Because of their relatively large drainage areas, which account for all of the crestline drainage from the Cascades into the Whidbey Basin, the Skagit and Snohomish rivers account for slightly more than half of the total freshwater discharged into all of Puget Sound. These are the rivers that have a principal input by snowmelt on the western flanks of the high Cascades. This results in two pronounced peaks in the flow rates through the year.

The Stillaguamish River is the third ranking river in terms of freshwater discharge. Of the three major rivers entering the Whidbey Basin, it is unique in that it does not directly drain to the crestline of the Cascades but lies between the Skagit subbasin (north) and Snohomish subbasin (south). The North Fork and South Fork converge near Arlington and the Stillaguamish is joined by Pilchuck Creek near Silvana. Flowing further west, the river enters Puget Sound at the north end of Port Susan.

Monthly averages of river discharge indicate peak flows in December–January and again in June. The winter peak is present over all of the Puget Sound region and is directly related to winter precipitation. The June peak in river discharge is the result of the spring snowmelt into the streams draining the upper western flanks of the Cascades. Some of the effect of snowmelt can be observed in the Main Basin, but the predominant effect is on the rivers entering the Whidbey Basin. The seasonal double peak in freshwater discharge into the whole of Puget Sound is principally the result of freshwater input from the rivers entering the Whidbey Basin. The least freshwater flow into Whidbey Basin and Puget Sound is during the summer months because the major spring snowmelt has ended and the summer months have the least amount of precipitation.

The Main Basin

The Main Basin is a poor second to the Whidbey Basin for freshwater input, accounting for less than 20 percent of the total input to the Puget Sound system. The rivers entering the Main Basin drain only a small part of the western slopes of the higher Cascades.

The Puyallup River is the largest of the rivers entering the Main Basin. It starts on the west slopes of Mount Rainier and is joined near Sumner by the White River, which originates on the north flanks of Mount Rainier. Flowing northwest from Sumner, the Puyallup enters Commencement Bay at the south end of the Main Basin north of The Narrows.

Two other principal rivers enter the Main Basin. The Green River originates near Stampede Pass, becomes the Duwamish at Orilla, and enters Elliott Bay at Seattle. The Sammamish and Cedar rivers enter

Lake Washington at its north and south ends respectively, and this fresh water ultimately enters the Main Basin at Shilshole Bay. Small additional sources of fresh water drain from Kitsap County along the western side of the Main Basin.

Highest freshwater discharge into the Main Basin occurs during December and January when more than 40 percent of the annual discharge takes place. Because of the smaller size of the Puyallup and Green river basins, the annual snowmelt in the higher slopes does not cause a pronounced secondary peak in annual flow cycle. Instead, high flow is maintained into the spring months, decreasing to the annual low flow rates of summer.

The Southern Basin

The Nisqually River is the fifth largest river entering the Puget Sound system and the principal one entering the Southern Basin. It accounts for almost half of the fresh water entering the system south of The Narrows. Originating on the south (southwest) flanks of Mount Rainier, the Nisqually is the only river entering the Southern Basin that has drainage from the upper western flanks of the Cascades and is the southernmost river originating in the Cascades to enter the Puget Sound system. Other sources of fresh water are the Deschutes River, which enters the south end of Budd Inlet at Olympia, minor drainage into Case and Carr Inlets off Kitsap Peninsula, and small streams entering Oakland Bay near Shelton and discharging into Hammersley Inlet.

All of the freshwater drainage into the Southern Basin accounts for only a little more than 10 percent of the total into Puget Sound. In proportion to its drainage area, freshwater discharge to the Southern Basin is disproportionately low compared with the Whidbey and Main basins. This is a result of several factors: the streams are shorter, annual precipitation is lower over the Southern Basin's watershed, and there is little snowmelt contributed from the higher precipitation areas of the Cascades. In addition, the entire Southern Basin drainage area is underlaid by thick, permeable, glacial deposits, which results in greater loss to groundwater than in other parts of the region.

The Hood Canal Basin

None of the principal rivers of the greater Puget Sound watershed enters Hood Canal, which receives only slightly more than 10 percent of the total fresh water entering Puget Sound. Part of this is provided by runoff in small streams from the Kitsap Peninsula and the southern part of Quimper Peninsula, but about two-thirds of the total originates in five small rivers that drain the eastern portion of the Olympic Peninsula. Of these, the Skokomish River is the largest and enters the Hood Canal near Union at the extreme southern end of the Great Bend. Do-

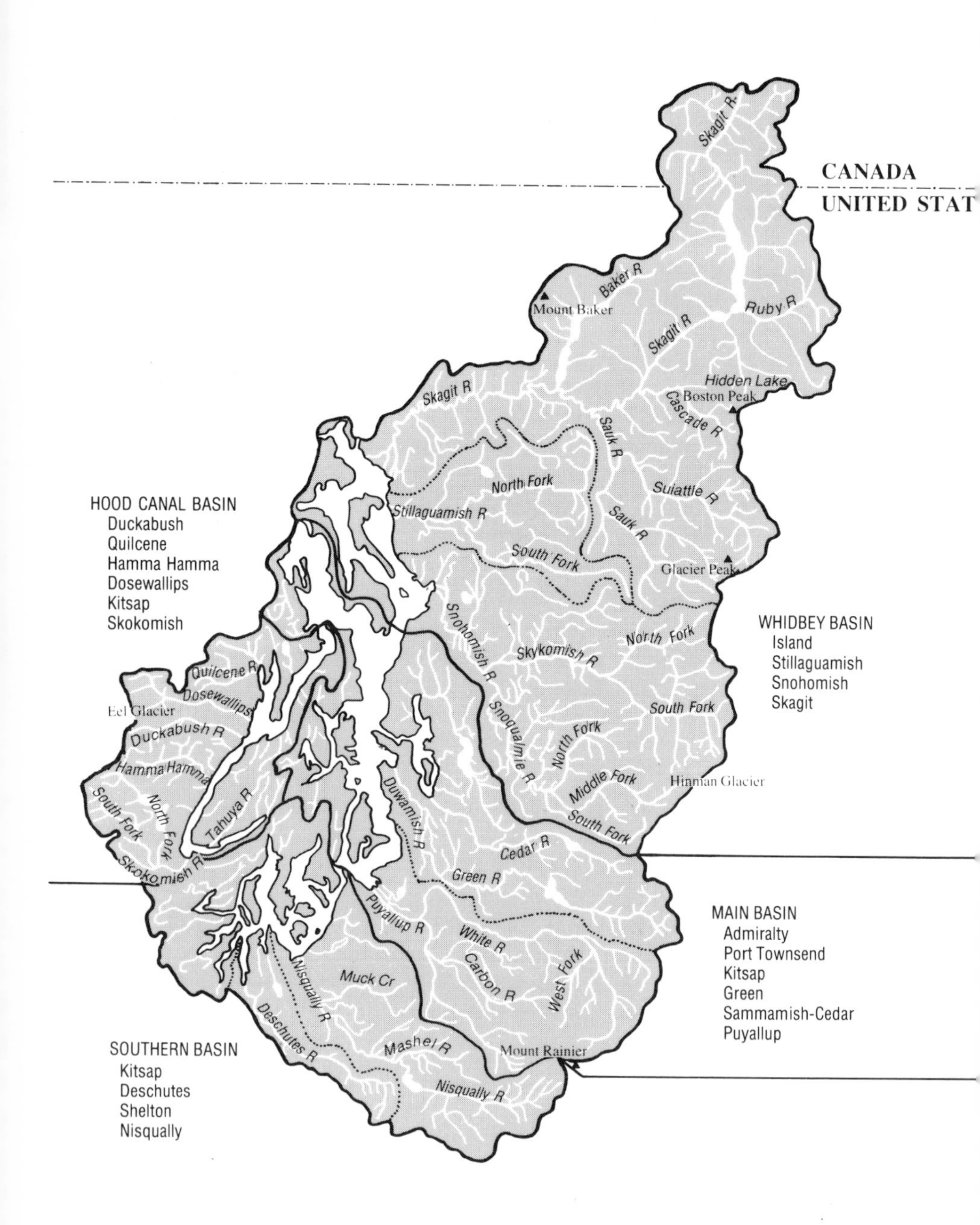

Figure 7.3 River drainage basins providing freshwater input to Puget Sound.

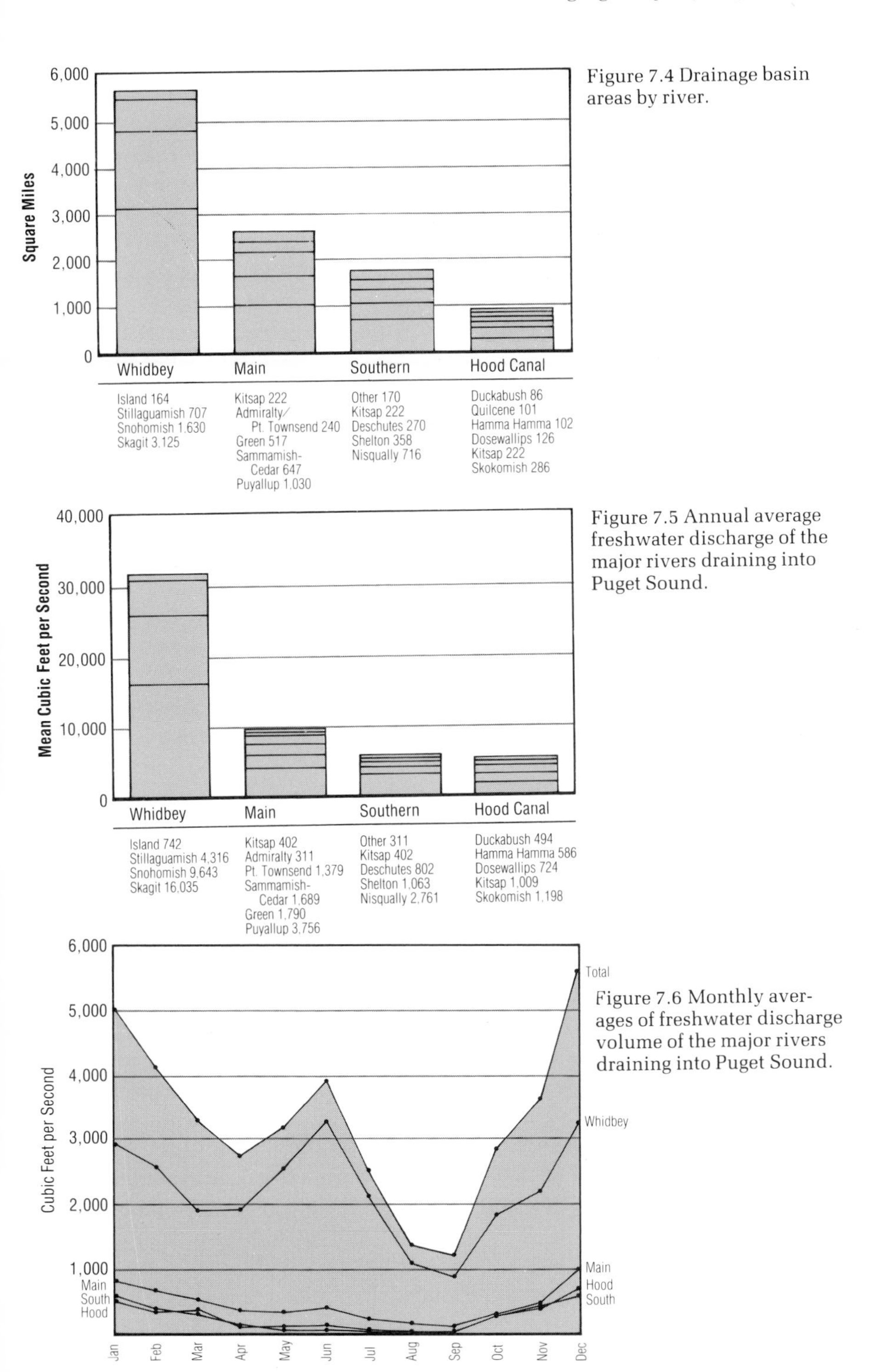

Figure 7.4 Drainage basin areas by river.

Figure 7.5 Annual average freshwater discharge of the major rivers draining into Puget Sound.

Figure 7.6 Monthly averages of freshwater discharge volume of the major rivers draining into Puget Sound.

sewallips, Duckabush, and Hamma Hamma are the sport fisherman's delight; and all pursue relatively short (30 kilometers; 20 miles) steep courses from elevations of 1,000–1,200 meters (3,000–4,000 feet) in the Olympics, and enter Hood Canal along its western side. The largest of these is the Dosewallips, which is fed by Eel Glacier at Mount Anderson. The only other prominent freshwater source is the the Quilcene River which enters Quilcene Bay at the town of Quilcene.

Seasonal distribution of freshwater discharge into Hood Canal follows the regional trend and is primarily a reflection of the annual precipitation cycles. There is some indication of a snowmelt peak in early spring since several of the streams originate in the Olympic Mountains where, principally because of lower elevations, snowmelt occurs earlier than in the Cascades.

Sediment Transport

Although some streams, particularly in the uplands, flow over bedrock, most flow over and through areas where there is an unconsolidated cover of soil or weathered earth. These unconsolidated fragments vary from large boulders to extremely small particles. Ongoing weathering reduces large fragments to small ones, and as the particles are transported, they are broken down further by abrasion and fragmentation.

Exactly how streams and other transporting agents carry the particles they move and what happens to the particles while they are being carried is extremely complex. Generally, particles are carried along a streambed by sliding or rolling (traction), are moved in a series of small jumps along and near the bed (saltation), or are suspended in the water (suspension). All of these processes are at work, in varying degree, in the rivers and streams of the Puget sound drainage basin.

The size and quantity of material a stream moves depends on how fast the water is moving and how large the cross-sectional area is. A stream's *competence* is a measure of the largest particle that can be moved: the faster the water is moving, the larger is the *largest* fragment that can be carried. The stream's *capacity* is a measure of the total volume of particles that the stream can carry. Capacity depends upon discharge, a measure of how much water is flowing in the stream. Since discharge is the product of speed multiplied by cross-sectional area, at comparable speeds larger streams carry more sediment than smaller ones. Generally, streams develop larger cross-sectional areas but move more slowly as water moves downward from the steep upland areas to broad flat riverbeds in the lowland. Consequently, although a river's capacity increases downstream, its competence usually decreases.

Although there is a general net movement of sedimentary particles downstream, this is usually accomplished in a series of steps for all but the very smallest grains. Larger rivers move a greater volume of sediment, but it tends to be finer material. In addition, seasonal variations of river discharge affect both competence and capacity. In high runoff periods, rivers move both larger particles and a greater total volume of material. During high runoff, rivers are competent to bring sand and even gravel-sized particles to the river mouth and into the water of Puget Sound.

Although the rivers that enter Puget Sound are the most obvious sediment source, almost an equal volume of particles are introduced into the Sound from non-point sources along the shoreline. This shoreline source derives its sediment from a variety of processes and introduces it along the shore rather than only at rivermouths. Shoreline introduced sediments are derived by natural processes such as beach and cliff erosion by wave action. In some cases, activities such as logging, beach protection, or dredging can result in considerable modification and, frequently, an increase in sediment available for shoreline transport and deposition.

Deposition in Puget Sound

In contrast with the high energy regimes of the rivers and the nearshore zone of wave action, the bulk of Puget Sound is a low energy environment. Lower energy is associated with decreased speed of water

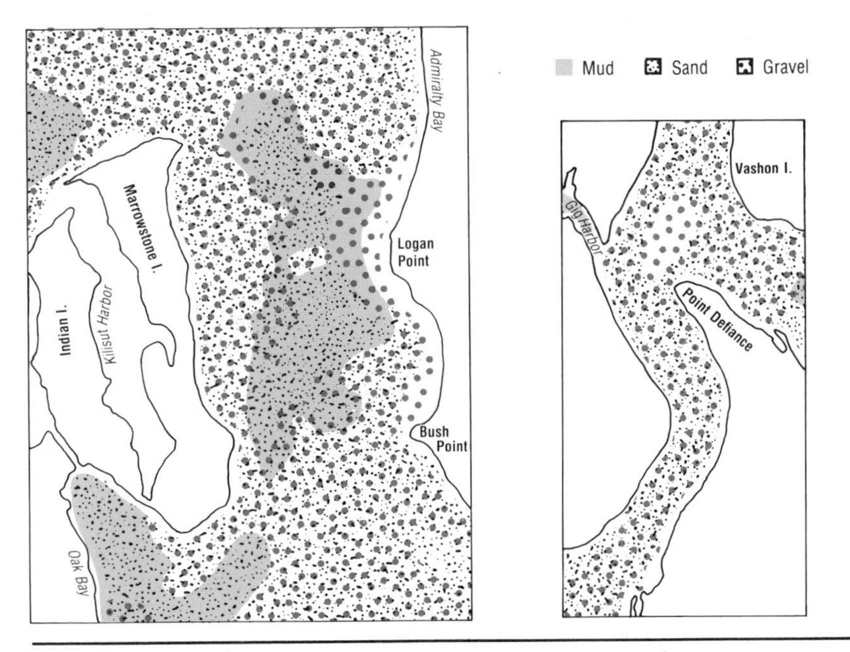

Figure 7.7 Patterns of gravel mixed with fine sediment occur in Admiralty Inlet and The Narrows. Sedimentation is not shown in waters shallower than ten meters.

movement, which results in increased deposition of sedimentary particles. Of all the sedimentary particles carried to the Sound, the great bulk are deposited within the Sound and only a very small percentage actually are carried out of the system beyond Admiralty Inlet.

Because it is covered by water, sedimentary material that has accumulated at the seafloor cannot be observed or identified directly. Although information concerning type and distribution of bottom sediments is required by geologists in order to answer many questions, all of these questions must be answered indirectly by examining seafloor samples brought to the surface for examination.

Sampling of bottom sediment may be conducted using submersibles or divers, but the most common methods involve collecting samples from a ship. Sampling programs vary depending on the information sought, the depth of the water, the nature of the bottom, and the analyses to be conducted.

Sampling devices have been designed to fit many different situations but belong to one of three general classes: dredges, grab samplers,

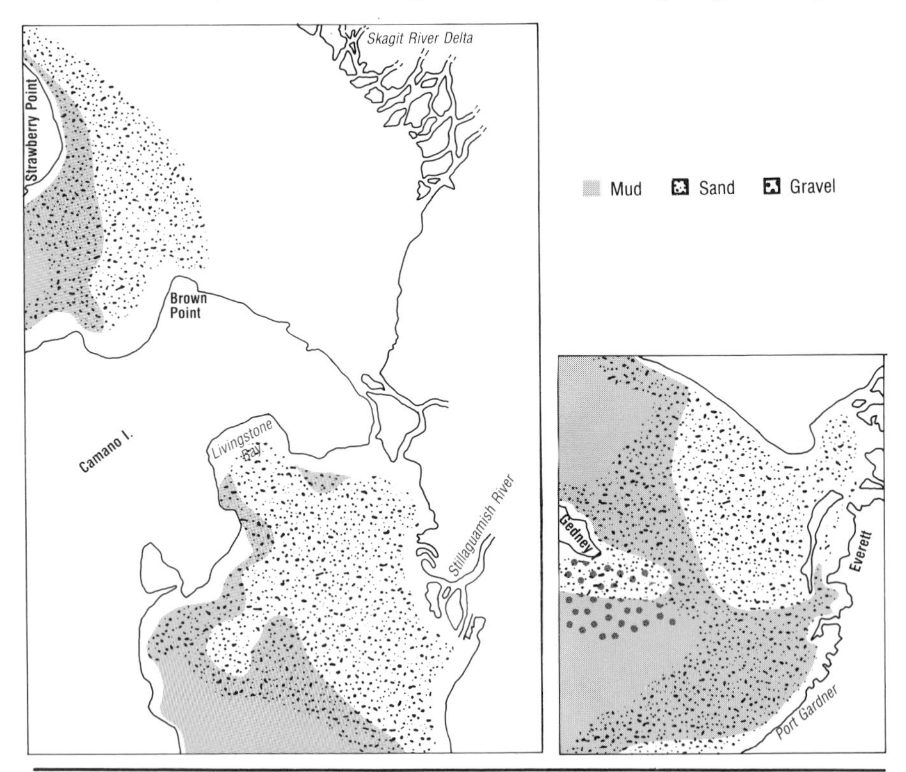

Figure 7.8 Where rivers deposit large volumes of sand and gravel, sandy deltas are formed, such as those at the mouths of the Skagit, Stillaguamish, and Snohomish rivers. Sedimentation is not shown in waters shallower than ten meters.

Figure 7.9 Where sediments are derived from shoreline sources rather than rivers, bands of sediment form parallel to the shore. Examples shown here are at Saratoga Passage and Case Inlet. Sedimentation is not shown in waters shallower than ten meters.

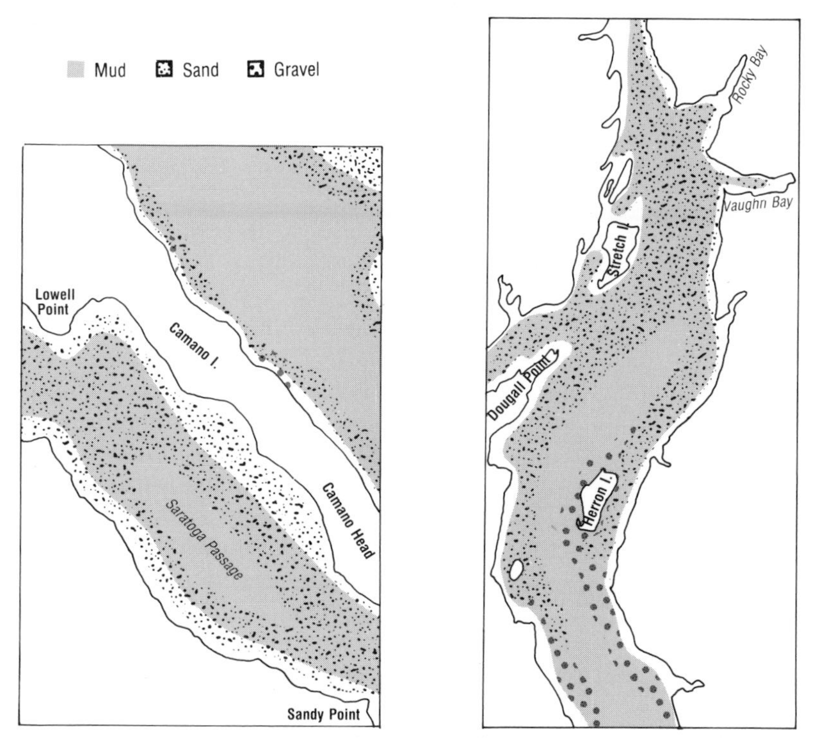

and coring devices. Ideally, the type selected depends on matching the characteristics of the sampling device and the type of information desired.

Variability in size, shape, and composition of sediments entering Puget Sound reflects the influence of weathering, erosion, and transportation by rivers and along the shore. Within the Sound, further distribution and eventual final deposition reflects interactions between the particles themselves and currents and wave action. The end product of all of this sedimentary activity is the slow filling of Puget Sound's basins. How fast this filling occurs differs depending on the rate of sediment input and the wave and current energy available to move them. Interaction between sediments and environmental energy results in distribution patterns in bottom sediment.

Although these patterns can be extremely diverse, some generalizations can be made that give insight into recent sedimentation and possible future evolution of the system.

Gravel mixed with fine sediment occurs where stronger than average currents remove much of the finer material from poorly sorted glacial drift, which originally covered the area. These deposits imply a net removal of sediment and relatively little deposition by postglacial sedimentation. Some of the more prominent examples of this condition are (Figure 7.7) Admiralty Inlet and The Narrows, as well as more localized occurrences such as Pickering Passage and Agate Pass.

A second characteristic bottom sediment distribution is the result of deposition of large volumes of sediment by rivers (Figure 7.8). Sand and gravel transported by flooding rivers accumulates close to the river mouth in the form of delta deposits. Subsequent sorting, resuspension, and removal of finer particles gives rise to the sandy deltaic deposits that grade away from the river mouths into finer bottom sediments. These deltaic deposits represent the most observable case of sedimentary "filling-in" in the region. The most prominent example of this sediment distribution is associated with the Skagit River, where deltaic deposits are rapidly filling in Skagit Bay. Other prominent examples are the Stillaguamish, Snohomish, and Nisqually river deltas.

A third characteristic bottom sediment distribution can be observed where the shoreline source of sediment overshadows the contribution of river inputs (Figure 7.9). Bottom sediment forms bands parallel to the shoreline with coarse sediment close to shore and fine material offshore. This gradation from coarse to fine is related to the gradation from faster to slower water motion (energy). In the shallow nearshore zone, the energy level is higher because wave action is more effective on the shallow bottom than in deeper water offshore. Nearshore, coarser materials may be moved and fine materials held in suspension. Offshore, very little coarse material is left to be moved and even the finest material tends to be deposited. This general distribution is probably the best approximation for a norm or average condition within the Sound; it represents a net filling in of the deeper portions of the Sound. Some more representative examples of this condition can be seen in Saratoga Passage, Carr and Case inlets, and Port Orchard.

What Next?

Because the shape of the surface of the earth's crust is continually changing, it is inevitable that Puget Sound will also change. The change will be the result of response to ongoing stresses applied by many of the same geological processes that have formed it. But the rate of change and the areas affected will remain comparable to what has happened in the past.

On the time-scale of a human lifetime we can expect to see changes resulting from sedimentation. The most obvious changes in the shape of Puget Sound are the result of sedimentary processes, both erosional and depositional. In many cases, human activities have impacted the natural sedimentation processes. Flooding and flood-control installation have changed the shape of the nearshore regions at the deltas of the Skagit, Snohomish, Duwamish, and Puyallup rivers. The sedimentary processes that have resulted in joining of Camano Island to the mainland, Maury to Vashon Island, and Marrowstone to Indian Island have all had a generous assist from the highway builders.

On a somewhat longer time-scale, further filling in of northeastern Skagit Bay can be expected to provide visible changes over the next several decades. The present sediment discharge into the Whidbey Basin is from the region's three largest rivers and results in the largest sediment accumulation rates for the entire Puget Sound system.

Most of the geological processes will work too slowly to be apparent to us or our immediate descendents. Although sedimentation is causing the most rapid change, it would take on the order of ten million years to completely fill in the Sound. It is highly improbable that Puget Sound would vanish, this way, however. Plate boundary processes are also acting and, though much more slowly, will continue to modify the shape of the crust—the Olympic and Cascade mountains as well as the Puget Lowland.

Over the next ten million years, changes should take place that will shift the plate boundary from convergent to shear and make the continental margin off Washington and Oregon more like California and southeastern Alaska. As the Gorda and Juan de Fuca spreading centers are progressively subducted, the offshore subduction zone and the associated convergent boundary processes will become increasingly less important. And on a time-scale of several tens of millions of years, there should be a relatively major reorientation of the global plates.

Glossary

accretion Addition of new material—e.g., fragments of continental crust at a subduction zone accreting to a larger section of continental crust.

alluvium Unconsolidated fragmental material deposited by streams in river beds, flood plains, lakes, fans at the foot of mountain slopes, and estuaries.

Andean-type boundary Plate boundary where oceanic crust and continental crust are converging; characterized by an offshore trench and a magmatic arc in nearby coastal mountains.

arc terrane An area characterized by a group of rocks or crustal material that originated on oceanic crust as an island arc.

asthenosphere A shell within the earth which is a zone of structural weakness capable of being plastically deformed; the shell below the lithosphere.

back-arc basin A basin, commonly underlain by oceanic crust, between an offshore volcanic island arc and a continent.

basalt An extrusive, fine-grained, dark igneous rock, rich in iron and magnesium; characteristic of oceanic crust.

basin A depression in the seafloor.

batholith A relatively large mass of igneous rock formed by deep-seated intrusions of magma.

bathymetry Measurement of the depth of a body of water; processed data depict the topography of the seafloor.

clastic Consisting of fragments of rock or of organic structures that have been moved individually from their places of origin to a place of deposition.

convergent boundary In plate tectonics, a boundary between two lithospheric plates that are converging; characteristically have a subduction zone where one of the plates moves under the other, and a magmatic arc on the overriding crust.

cordillera A group of mountain ranges, basically parallel, belonging to a single mountain building system.

crust Solid outer layer of the earth.

detritus Fine, particulate debris, usually of inorganic material produced by the weathering and disintegration of rock.

divergent boundary A boundary between two lithospheric plates that are moving apart, characteristically with new oceanic crust being created along a submarine ridge.

echo sounding Means of measuring the depth of water by measuring the time interval between the release of a sound pulse and the return of its echo from the bottom.

epicenter The point on the earth's surface directly above the focus of an earthquake.

erratic A transported rock fragment different from the bedrock on which it lies; generally applied to fragments transported by glacial ice or floating ice.

extrusive rock Those igneous rocks derived from magma poured out or ejected at the earth's surface; synonymous with volcanic.

focus Seismic focus; the point in the earth's crust where an earthquake shock originates.

foliation Layering or lamination in metamorphic rock.

fracture zone Long, linear zone of irregular topography on the seafloor, frequently associated with shear boundaries between lithospheric plates.

geomagnetic field The natural magnetic field of the earth.

glacial drift Sediment in transport by glaciers, deposited by glaciers, or predominantly of glacial origin.

glaciofluvial Pertaining to streams flowing from glaciers or the deposits made by such streams.

glaciolacustrine Pertaining to glacial meltwater lakes or deposits accumulated within such lakes.

igneous rock Rock formed by cooling and solidifying of initially molten magma.

interstade Short period of time within a longer period of glaciation when the terminus of the glacial ice is retreating.

intrusive rock Igneous rocks which originate when intruded magmas cool at depth without reaching the surface.

island arc Chain of volcanic islands formed on oceanic crust along a magmatic arc when two oceanic plates converge.

Japan-type boundary Convergent boundary characterized by an offshore magmatic arc separated from the continent by an oceanic basin.

kame A hill or short irregular ridge of gravel or sand deposited from meltwater in contact with glacial ice.

kettle A depression in drift made by the wasting away of a detached mass of glacier ice that had been either wholly or partly buried in the drift.

lacustrine clay Clay produced by or formed in a lake.

lithosphere The outer, rigid portion of the earth; includes the continental and oceanic crust and the upper part of the mantle.

mafic rock Rock that is rich in iron and magnesium; generally synonymous with "dark minerals."

magma Molten rock material that forms igneous rocks upon cooling. Magma that reaches the earth's surface is referred to as lava.

metamorphic rock Includes all those rocks which, although already solid, have been changed in texture, composition, or structure in response to pronounced changes of temperature, pressure, and chemical environment.

mid-ocean ridge An elongated mountain range rising from the ocean floor and found extending through the North and South Atlantic Oceans, the Indian Ocean and the South Pacific Ocean.

mineralogy The study of minerals.

moraine Glacial deposit of rock, gravel, and other sediment.

nunatak A hill or peak which was formerly surrounded but not overridden by glacial ice.

orogeny The process of forming mountains, particularly by folding and thrusting.

oxides Materials resulting from chemical combination with oxygen.

pelagic Related to the seawater, and to organisms and inorganic materials that originate or are found in seawater.

petrology The study of rocks.

piedmont glacier A mass of glacial ice formed by coalescence of several valley glaciers.

pillow lavas Lava that has solidified under water in the appearance of a pile of pillows.

saltation Movement of sediment particles in running water in a series of small jumps along the streambed.
sedimentary rock Rocks that result from accumulation of fragments and dissolved material.
sedimentation Process involving transportation and deposition of particulate materials and associated dissolved material.
seismology The study of earthquakes.
shear boundary In plate tectonics, a boundary between two adjoining lithospheric plates where the plates move laterally.
shoal A shallow area in a river, sea, or lake; elevation of the sea bottom.
silicic rock Rock that is rich in silica, commonly with quartz (silica dioxide); generally lighter colored rocks.
sill Shallow area of the seafloor that separates two basins from one another or a coastal bay from the adjacent ocean.
stade Period of glaciation during which the terminus of the ice advances.
stream capacity Measures the total load a stream can carry.
stream competence Measures the stream's ability to move a particle of given size; indicated by the size of the largest particle that can be moved.
subaerial Occurring on the surface (subatmospheric) of the earth, as contrasted with "submarine."
subduction zone Zone where one of two converging plates moves under the other; commonly associated with deep sea trenches.
superposition The process by which something (e.g., mudflow, lava, sediment) is laid down upon an existing surface and is therefore younger than the surface it covers.
till Unsorted, heterogeneous materials deposited directly by glaciers without subsequent transport by water.
topography The general shape and patterns of elevation of land surface or the ocean bottom.
trench Long, deep, and narrow depression of the seafloor with relatively steep sides, associated with a subduction zone.
valley glacier A mass of ice moving downward and outward from an upland field of accumulating ice, commonly moving along a pre-existing valley.
weathering The disintegration and decay of rock producing an accumulation of altered material; the process occurring when rocks at the earth's surface interact principally with water and changing temperature.

Bibliography

Anderson, F. E. 1968. Seaward terminus of the Vashon Continental Glacier in the Strait of Juan De Fuca. *Marine Geology*. 6:419–438.

Armstrong, J. E., D. R. Crandell, D. L. Easterbrook, and J. B. Noble. 1965. Late Pleistocene stratigraphy and chronology in southwestern British Columbia and northwest Washington. *Geological Society of America Bulletin*. 76:321–330.

Atwater, T. 1970. Implications of plate tectonics for the Cenozoic tectonic evolution of western North America. *Geological Society of America Bulletin*. 81:3513–3536.

Barr, S. M. 1974. Structure and tectonics of the continental slope west of southern Vancouver Island. *Canadian Journal of Earth Sciences*. 11:1187–1199.

Barr, S. M. and R. L. Chase. 1974. Geology of the northern end of Juan de Fuca Ridge and seafloor spreading. *Canadian Journal of Earth Sciences*. 11:1384–1406.

Bullard, F. M. 1962. *Volcanoes: In History, In Theory, In Eruption*. Austin: University of Texas Press. 441 pp.

Campbell, C. D. 1953. Washington geology and resources. *Research Studies of the State College of Washington*. XXI(2):114–153.

Cotton, C. A. 1952. *Volcanoes as Landscape Forms*, Second Edition. New York: John Wiley & Sons, Inc. 416 pp.

Crandall, D.R. 1965. The glacial history of western Washington and Oregon. In: H. E. Wright, Jr. and D. G. Frey, eds., *The Quaternary of the United States*. Princeton, New Jersey: Princeton University Press. Pp. 341–353.

Crandall, D. R., D. R. Mullineaux, and H. H. Waldron. 1958. Pleistocene sequence in southeastern part of the Puget Sound lowland, Washington. *American Journal of Science*. 256:384–397.

Curray, J. R. 1965. Late quaternary history, continental shelves of the United States. In: H. E. Wright, Jr. and D. G. Frey, eds., *The Quaternary of the United States*. Princeton, New Jersey: Princeton University Press. Pp. 723–735.

Danner, Wilbert R. 1955. *Geology of the Olympic National Park*. Seattle, Washington: University of Washington Press. 68 pp.

Dehlinger, P., R. W. Couch, D. A. McManus, and M. Gemperle. 1971. Northeast Pacific structure. In: A. E. Maxwell, ed., *The Sea*, Vol. 4. New York: John Wiley & Sons. Pp. 133–189.

Easterbrook, D. J. 1969. Pleistocene chronology of the Puget Lowland and San Juan Islands, Washington. *Geological Society of America Bulletin*. 80:2273–2286.

Easterbrook, Don J. and David A. Rahm. 1970. *Landforms of Washington*. Bellingham: Western Washington State College.

Ekman, Leonard C. 1962. *Scenic Geology of the Pacific Northwest*. Portland, Oregon: Binfords and Mort, Publishers.

Elvers, D., S. P. Srivastava, K. Potter, J. Morley, and D. Sdidel. 1973. Asymmetric spreading across the Juan de Fuca and Gorda Rises as obtained by a detailed magnetic survey. *Earth Planet Sciences Letters*. 20:212–219.

Freeman, O. W. and H. H. Martin. 1954. *The Pacific Northwest*. New York: John Wiley & Sons.

Friebertshauser, M. A. and A. C. Duxbury. 1972. A water budget study of Puget Sound and its subregions. *Limnology and Oceanography*. 17(2):237–247.

Harris, Stephen. 1980. *Fire and Ice*. Seattle, Washington: The Mountaineers and Pacific Search Press.

Hawkins, N. M. and R. S. Crosson. 1975. Causes, characteristics, and effects of Puget Sound earthquakes. *Proceedings of the U.S. National Conference on Earthquake Engineering*. 1975:104–112.

Highsmith, R. M., ed. 1973. *Atlas of the Pacific Northwest*, Fifth Edition. Corvallis, Oregon: Oregon State University Press. 128 pp.

King, Philip B. 1977. *The Evolution of North America*, Revised Edition. Princeton, New Jersey: Princeton University Press. 197 pp.

Kulm, L. D. and G. A. Fowler. 1974. Cenozoic sedimentary framework of the Gorda-Juan de Fuca plate and adjacent continental margin: a review. In: *Modern and Ancient Geosynclinal Sedimentation; Deposits in Magmatic Arc and Trench Systems*. Society of Econ. Paleontology and Mineralogy, Special Paper 19, pp. 212-229.

Lincoln, J. L. 1977. Derivation of freshwater inflow into Puget Sound. University of Washington, Dept. of Oceanography, Special Report No. 72. Reference M77-28, March 1977, 20 pp.

McKee, B. 1972. *Cascadia: The Geologic Evolution of the Pacific Northwest*. New York: McGraw-Hill Book Company. 394 pp.

McWilliams, R. G. 1978. Early tertiary rifting in western Oregon-Washington. *American Association of Petroleum Geologists Bulletin*. 62(7):1193–1197.

Mackin, J. H. and A. S. Cary. 1965. Origin of Cascades landscapes. Washington Division of Mines and Geology, Information Circular 41, 35 pp.

Matthews, William H. 1968. *Guide to the National Parks, Their Landscape and Geology*, Volume I, Western Parks. New York: Natural History Press.

Mayers, I. R. and L. C. Bennett, Jr. 1973. Geology of the Strait of Juan de Fuca. *Marine Geology*. 15:89–117.

Peattie, Roderick, ed. 1949. *The Cascades*. New York: The Vanguard Press.

Riddihough, R. P. 1972. A model for recent plate interactions off Canada's west coast. *Canadian Journal of Environmental Science*. 14:384-396.

————————. 1978. The Juan de Fuca Plate. *Transactions, American Geophysical Union*. 59(9):836–842.

Riddihough, R. P. and R. D. Hyndman. 1976. Canada's active western margin—the case for subduction. *Geosciences Canada*. 3:269–278.

Rona, P. A. and E. S. Richardson. 1978. Early cenozoic global plate reorganization. *Earth and Planetary Science Letters*. 40:1–11.

Silver, E. A. 1971. Small plate tectonics in the northeastern Pacific. *Geological Society of America Bulletin*. 82:3491–3496.

———————. 1972. Pleistocene tectonic accretion of the continental slope off Washington. *Marine Geology*. 13:239–249.

———————. 1978. Geophysical studies and tectonic development of the continental margin off the western United States, lat. 34° to 48°N. In: R. B. Smith and G. P. Eaton, Eds., *Cenozoic Tectonics and Regional Geophysics of the Western Cordillera*. Geological Society of America Memoir 152, pp. 251–262.

Simpson, R. W. and A. Cox. 1977. Paleomagnetic evidence for tectonic rotation of the Oregon coast range. *Geology*. 5:585–589.

Snavely, P. D. and H. C. Wagner. 1963. Tertiary geologic history of western Oregon and Washington. Washington Division of Mines and Geology Report. Inv. 22, 25 pp.

Stearn, C. W., R. L. Carroll, and T. H. Clark. 1979. *Geological Evolution of North America*, third edition. New York: John Wiley and Sons. 566 pp.

Tabor, R. W. and W. M. Cady. 1978. The structure of the Olympic Mountains, Washington—analysis of a subduction zone. U.S. Geological Survey Professional Paper 3033. Washington, D.C.: U. S. Government Printing Office. 25 pp.

Thorson, Robert M. 1980. Ice—sheet glaciation of the Puget Lowland, Washington, during the Vashon stade (late Pleistocene). *Quaternary Research*. 13(3):303–321.

Tiffin, D. L., B. E. B. Cameron, and J. W. Murray. 1972. Tectonic and depositional history of the continental margin off Vancouver Island, B. C. *Canadian Journal of Earth Sciences*. 9:280–296.

Tobin, D. G. and L. R. Sykes. 1968. Seismicity and tectonics of the northeast Pacific Ocean. *Journal of Geophysics Research*. 73:3821–3845.

U. S. Geological Survey. 1978. Water Resources Data for Washington; Water Year 1977; Vol. 1, Western Washington. U. S. Geological Survey Water-Data Report WA-77-1, Tacoma, Washington, 429 pp.

Weissenborn, A., ed., and F. W. Cater. 1966. The Cascade Mountains. In: *U. S. Geological Survey, Mineral and Water Resources of Washington*, Division of Mines and Geology, Reprint 9, pp. 27–37.

Index

Other books in the Puget Sound Series

The Water Link
A History of Puget Sound as a Resource
Daniel Jack Chasan

Governing Puget Sound
Robert L. Bish

Marine Birds and Mammals of Puget Sound
Tony Angell and Kenneth C. Balcomb III

The Coast of Puget Sound
Its Processes and Development
John Downing

The Fertile Fjord
Plankton in Puget Sound
Richard M. Strickland